Introduction to Mass Spectrometry

Introduction to Mass Spectrometry

by H. C. Hill
formerly Imperial Chemical Industries Ltd.,
Heavy Organic Chemicals Division, Billingham

SECOND EDITION

revised by A. G. Loudon
Department of Chemistry,
University College, London

HEYDEN & SON LTD
London · New York · Rheine

Heyden & Son Ltd., Spectrum House, Alderton Crescent, London NW4 3XX.
Heyden & Son Inc., 225 Park Avenue, New York, N.Y. 10017, U.S.A.
Heyden & Son GmbH, 4440 Rheine/Westf., Münsterstrasse 22, Germany.

© Heyden & Son Ltd., 1972.

All Rights Reserved. No part of this publication may be reproduced, stored in a retrieval system, or transmitted, in any form or by any means, electronic, mechanical, photocopying, recording or otherwise, without the prior permission of Heyden & Son Ltd.

Library of Congress Catalog Card No. 66-19670

ISBN 0 85501 038 X (paperback)
ISBN 0 85501 039 8 (cloth)

First Edition 1966
Reprinted 1969
Second Edition 1972

Printed in Great Britain by William Clowes & Sons, Limited, London, Beccles and Colchester.

Contents

Preface		vii
Chapter 1	Instrumentation and Sample Handling	1
	Introduction	1
	Historical development	1
	Applications	1
	Recording of spectra	2
Chapter 2	Basic Aspects of Organic Mass Spectrometry	10
	Processing of a mass spectrum	10
	Important spectral features	14
	Other techniques	23
Chapter 3	Fragmentation of Positive Ions	35
	Introduction	35
	Simple fission processes	39
	Rearrangement processes	48
	Complex fission processes	62
	Formation of a mass spectrum	67
Chapter 4	Interpretation of the Mass Spectrum	70
	Introduction	70
	Summaries of fragmentation processes applicable to particular types of compounds	70
	An approach to the determination of the structure of a molecule with the aid of mass spectrometry	91
	Characteristic fragmentation patterns	93
	The mass spectrometric shift technique	96
	Examples of structural determination using mass spectrometry	96
	Qualitative and quantitative analysis of mixtures	106
	Application of mass spectrometry to problems in organic chemistry	107
	Sources of information and data regarding mass spectrometry	107
	Index to unknown spectra	111
References		112
Index		113

Preface to Second Edition

The first edition of this book was widely used in mass spectrometry courses throughout the world; the recent reprint and the book's translation into Japanese, Italian and German gives some idea of this demand. It was therefore the more tragic that the author did not live to see his book acclaimed — Hilson C. Hill died in 1967, one year after the book was first published.

In revising this book I have tried to keep, as far as possible, to the original style and aims of the late author. The recent advances in this field, particularly in instrumentation, have, however, made major changes necessary in some portions and the publishers have taken advantage of this opportunity to completely reset the text in a more readable form.

March, 1972   Alec G. Loudon

Preface to First Edition

During the last few years mass spectrometry has been rapidly developed and has become an indispensable tool for the determination of the structure of organic molecules; the method should preferably be used in conjunction with the other spectroscopic techniques.

For several years most undergraduate courses have included lectures on spectroscopy, and graduate chemists appreciate the applicability of infrared, ultraviolet and nuclear magnetic resonance spectroscopy to structure elucidation problems. This book has been written to give organic chemists a similar understanding of the potential of mass spectrometry and to enable them to interpret a mass spectrum as completely as is necessary.

In outlining the mechanistic approach used to rationalize the fragmentation of the positive ions formed from organic molecules in a mass spectrometer, an attempt has been made to incorporate all existing outlooks, so that when other textbooks or papers are consulted the arguments put forward will be familiar.

The existing textbooks of Professors Biemann, Djerassi and McLafferty present these outlooks in great detail and have been consulted in the preparation of the discussion of molecular ion fragmentation presented in Chapter 3.

The molecules discussed and the examples used have been kept simple in order to impress upon the reader that the fragmentation mechanisms so far evolved explain the genesis of the major peaks in the mass spectra of a very wide range of organic compounds. When these mechanisms have been memorized they can be employed to rationalize the fragmentation behaviour of the molecular ions formed from complicated molecules.

Research is in progress in organic mass spectrometry and although some of the mechanisms may later prove to be faulty or require modification it is better to employ these until something better replaces them, or this approach to mass spectrometry becomes obsolete.

Some of the more elaborate mechanisms, some topics of mass spectrometry and some classes of compounds have not been discussed in this book which was not intended to be fully comprehensive. The reader is referred where appropriate to textbooks which are more detailed and which themselves refer to the original papers.

I gratefully acknowledge the advice and useful criticism of Dr. A. E. Brown of this Department, and Dr. W. Kelly of Unilever Research Laboratories, who gave up a lot of their time to read and comment on the entire manuscript. I would also like to thank Professor A. Maccoll of University College, London, who read and advised on the typescript and Mr. Martin Elliott of A. E. I. Ltd., Manchester who supplied advice and data for the instrumentation chapter, and with his Company's approval, diagrams of their instruments and sample handling systems.

I also extend my thanks to Professor C. Djerassi, Professor K. Biemann and the McGraw-Hill Book Company, and Dr. J. H. Beynon and the Elsevier Publishing Company for permission to make use of copyright material, and to the Dow Chemical Company for permission to make use of mass spectra from their published collection.

Finally, I am indebted to all the mass spectroscopists whose research, although not directly referred to in this text, has enabled mass spectrometry to achieve its present importance and has provided the information which has made the writing of this book possible.

May, 1966　　　　　　　　　　　　　　　　　　　　　　　　　　　　　　　　　　Hilson C. Hill

1
Instrumentation and sample handling

INTRODUCTION
A mass spectrometer is an instrument in which ions are produced from a sample, separated according to their mass to charge ratios, and recorded.

Ionization is a process whereby an electrically neutral molecule becomes positively or negatively charged.

HISTORICAL DEVELOPMENT
In 1898 Wien showed that a beam of positive ions could be deflected using electric and magnetic fields. In 1912 Thomson showed the existence of two neon isotopes using a simple magnetic deflection instrument. In 1918 Dempster and in 1919 Aston designed more elaborate instruments which were used for the measurement of the relative abundances of isotopes. Reliable mass spectrometers for use in the petroleum industry became available during the early 1940's. These could be operated by chemists without any training in electronics. Several methods of sample handling have been developed and performance has been steadily improved until most organic molecules, up to approximately fifteen hundred in molecular weight and of melting point below 350°C, can be studied using a single focusing instrument. The double focusing instrument has been designed to improve the power to separate ions of different mass (resolving power). This makes possible measurement of the mass of an ion to an accuracy of within a few parts per million, thus yielding its atomic constitution.

APPLICATIONS
Mass spectrometers are used for a wide variety of purposes in many branches of science involving the study of atoms and molecules. The nature of the mass spectrometer varies according to the nature of the study, but the basic processes — sample preparation and introduction, ionization, separation of the ions according to their mass to charge (m/e) ratios, collection and recording of the ions — are common to all applications.

Chemists employ mass spectrometers in three ways which are, to some extent, related or inter-dependent.
1. The study of the behaviour of molecules under electron impact and the behaviour of the ions produced.
2. Determination of the structure of molecules.
3. The qualitative and quantitative determination of the components of a mixture.

In the case of metals and alloys somewhat special techniques have to be used and these will not be discussed in this book.

RECORDING OF SPECTRA

As mentioned above there are essentially four steps in the recording of a spectrum. In all cases the apparatus runs at a high vacuum and this is implicit in the following discussion. It does not matter in principle whether the ions are positive or negative, but most of the work has been done on positive ions. The essential points of the four steps are discussed in the next sections.

Sample preparation and introduction

Normally the sample must be in the vapour state before ionization and three general methods of achieving this have been developed.

Reservoir system

After the removal of any air, gases and very volatile liquids with a vapour pressure of greater than approximately 10^{-3} torr at room temperature are admitted into a reservoir. This reservoir will typically have a volume of about 1 litre and work at a sample pressure up to 10^{-2} torr or so. The gas is admitted into the ionization region from the reservoir via a line fitted with a leak. This system maintains a steady pressure of between 10^{-6} and 10^{-5} torr in the source, the latter being the region of the mass spectrometer containing the ionization apparatus (see Fig. 1).

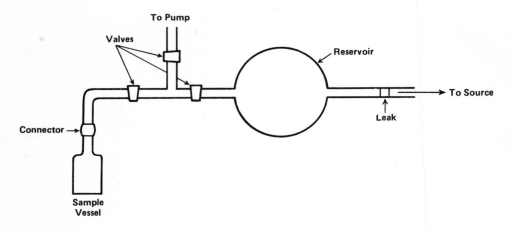

FIG. 1. Reservoir system for introduction of gases and volatile liquids.

When the sample is less volatile it is distilled into a heated reservoir which is connected in a similar way to the source, although in this case the line will also be heated. To do this the sample can either be placed in a small tube, frozen, the air removed by pumping and the sample then distilled, or a gallium inlet system can be used. In the latter case the sample is drawn into a pipette, which is pushed through a layer of molten gallium ($t = 200°C$) to make contact with a sintered disc. Due to the temperature of the gallium and the high vacuum the sample will distil into the heated reservoir of the machine. In both cases the sample may be pyrolysed by the procedure, particularly if there are hot metal parts in the reservoir or inlet line. An all-glass inlet system is preferable for this reason.

Direct insertion system
This was first developed for use with very involatile liquids and solids. Although some solids can be introduced using the heated inlet system described above, in many cases this is either impossible or leads to a spectrum of the pyrolysis products. This method was developed to deal with these problems.

A small sample is placed in a suitable holder and the latter inserted via a vacuum lock to a position near the ionizing region in the source. Either the heat of the source (see p. 4) or, preferably, a separate heating system then produces a sufficiently high vapour pressure of the compound to give a large enough ion current from which a spectrum can be obtained. The separate heating system is preferable because the metal source can be cooled and hence pyrolysis avoided. Originally this method was restricted to less volatile compounds, because the source pressure would have been too high. However, with a separately heated or, indeed, cooled direct insertion system volatile substances can be introduced in this manner. One minor disadvantage of this method is that it is not always so easy to produce a constant sample supply.

Direct sampling from GC machines
Although it is, of course, possible to trap each fraction from a gas liquid chromatography (GC) machine and introduce it into the mass spectrometer by one of the methods described above, it is normally much more convenient to determine the spectrum of each component of the mixture as it is eluted from the column. To do this the exit of the column is connected via a separator to the source of the mass spectrometer. The separator preferentially removes the carrier gas, often helium, from the mixture of gas and compound, in order that the total pressure in the source be acceptable ($\sim 10^{-5}$ torr). The inlet to the source is so constructed that, after separation, the enriched sample is directed into the ionization region (see Fig. 2). Since a typical gas chromatography fraction is eluted in less than a minute this technique requires a fairly fast recording of the spectrum.

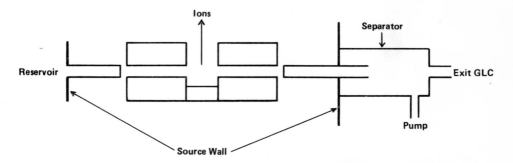

FIG. 2. GC inlet apparatus.

Sample size
One of the advantages of mass spectrometry compared with other analytical methods is the small size of sample which is required. Some typical and minimum sample sizes required for the above systems are shown in Table 1.

TABLE 1. Sample sizes for various inlet systems.

System	Typical sample	Minimum sample
Cold inlet	1 std. ml	10^{-5} std. ml
Heated inlet	100 µg	1 µg
Gallium inlet	100 µg	1 µg
Direct insertion (probe)	20 µg	0.1 µg
GC sampling	1 µg	0.01 µg

Samples should be submitted for examination in containers which allow easy transfer to one or other of the sampling systems described. Ampoules should be capable of being broken open easily; small amounts of liquid should be in a small tube with a hemispherical bottom so that the liquid accumulates and can easily be removed; gases should be in a container fitted with a vacuum stopcock and a standard taper socket.

Ionization

There are four important methods in use at the moment:

Electron bombardment

This is the most widely used method. Electrons of a fixed energy between 5 and 100 eV (1 eV = 23.05 kcal) are allowed to collide with the vapour of the substance under investigation. This method produces not only molecular ions but also, in abundance, fragment ions. Of the methods described here this one produces the largest amount of fragmentation which makes it particularly useful for structure determination, but sometimes the molecular ion is not detectable. It also has the advantage of producing the highest total ion current and is the only method which allows a direct study of the energetics of the formation of a particular ion. The electrons are produced by heating a filament and the heat so generated is responsible for keeping the source warm, which, as mentioned earlier, may cause pyrolysis.

Chemical ionization

The sample is admitted into the source in the appropriate way with a large excess of methane or some other simple hydrocarbon gas and the mixture bombarded by electrons. Since ionization by electron bombardment is itself a relatively inefficient process (typically in a mass spectrometer about 1 per cent of the molecules are ionized) this bombardment results mainly in the ionization of the hydrocarbon. These ions in turn ionize the sample by charge exchange. This results in a relatively small transfer of energy to the sample (*cf.* electron impact) and thus the molecular ion tends to be more intense relative to the fragment ions than in an electron bombardment spectrum of the normal type. Since very little fragmentation occurs, this technique is less suitable for structural determination. Often, ion—molecule reactions occur which can result in the production of ions with m/e values higher than that of the molecular ion, and this has to be remembered when trying to identify the latter.

Field ionization

The sample is passed through a very strong electrical field which ionizes it. The field also acts as the accelerating potential required for the ion separation. Such a field is normally generated by applying a high potential (approximately 10 kV) to a thin wire or razor blade. This method produces mainly molecular ions, but ion—molecule reactions give rise to dimer and trimer molecular ions, which has to be borne in mind when looking for the molecular ion. This method can also be applied to the study of the rates of reaction of ions.

Photo-ionization

The sample vapour is irradiated with light of a sufficiently high energy (i.e. frequency since $E = h\nu$) to cause ionization. Often the helium line at 584 Å ($E = 21$ eV) is used. This has the advantage that the incident energy is known accurately, which is not the case in any of the other methods described. By analysis of the energy of the electrons given off, the ionization potentials of molecules (see p. 28) can be determined more exactly.

Ion separation

The ions are separated according to their m/e values. They are first accelerated through a slit (the source slit, see p. 7) by passing them through a high potential. In *Field ionization* above the ions are automatically accelerated by the ionization process. After this every mono-charged ion has, to a first approximation, the same energy. After acceleration the following methods of separation are the most common.

Use of magnetic fields

This is the most common method and two different general types of design are used. Both of these depend on the fact that a magnetic field will deflect a moving charged particle. The fundamental physical theory is very similar to that applied to determine the force exerted by a magnet on a wire carrying a current.

Single focusing instruments. This design is used in mass spectrometers employed for the routine production of low resolution mass spectra (see p. 10) although in some cases accurate mass measurements can be made. It can be shown that the equation governing the flight of a charged particle is given by:

$$m/e = H^2 R^2 / 2E \tag{1}$$

This is derived as follows:

After acceleration by passing through the accelerating field (E) an ion with a charge (e) has kinetic energy (eE) and this can be equated to the usual expression for kinetic energy:

$$eE = \tfrac{1}{2} m v^2 \tag{2}$$

The magnetic field (H) forces the charged particle to move in a circle of radius R. Thus the force exerted by the magnet on the ion, which is Hev, must be equated to the centrifugal force which is mv^2/R.

$$Hev = mv^2/R \tag{3}$$

Eliminating v between Eqns. (2) and (3) gives Eqn. (1) above. This derivation assumes that all the ions of the same formula and charge have the same energy, which is only true to a first approximation. Inspection of Eqn. (1) shows that ions of different m/e can be separated by varying R, H or E.

One way of recording the spectrum is with a photoplate placed in the appropriate position, this being equivalent to varying R. Alternatively, a collector can be placed in one position (i.e. R fixed) and by scanning either the accelerating voltage (E) or the magnetic field (H) each type of ion (i.e. ions of the same m/e value) can be brought to focus in turn

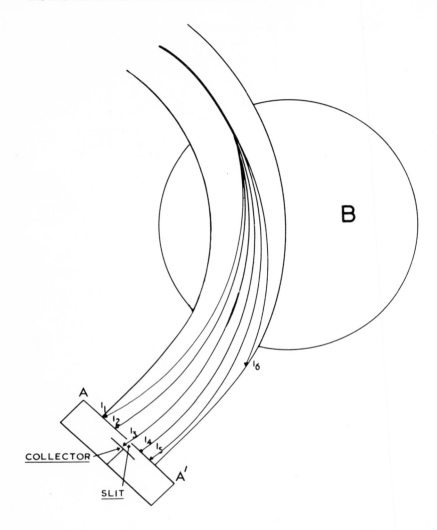

FIG. 3. Principle of operation of a single focusing mass spectrometer. $I_1 - I_6$ represent beams of ions of one m/e value and B represents the magnet.

on this and recorded. In the case of combined GC/MS, it is normal to scan the accelerating voltage to obtain the necessary scan speed. However, if the accelerating voltage is altered this results in a discrimination against the ions of high m/e value, so that the intensities of the signals are not proportional to the relative intensities of the ions.

The principle of the working of this type of mass spectrometer is shown above (Fig. 3) for the case of a collector. In the case of a photographic plate it would be placed along the surface A–A' and a collector slit of appropriate dimensions would be placed in front of this. The general design of the instrument is similar to that for a double focusing instrument (see below) but without the electrostatic analyser.

Double focusing instruments. These were designed to account for the fact that the ions after acceleration have, in practice, a finite range of kinetic energies, not a unique kinetic energy as assumed in the previous section. This range of kinetic energies leads to peak broadening and eventually to some, or complete, overlapping of peaks corresponding to ions of very similar masses (see p. 24).

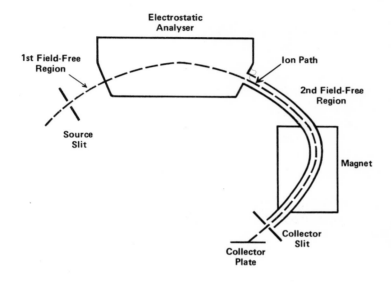

FIG. 4. Nier–Johnson geometry.

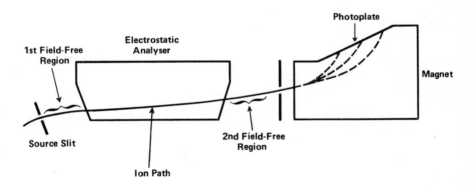

FIG. 5. Mattauch–Herzog geometry.

To overcome this, an electrostatic analyser is inserted between the source and the magnet. This consists of a chamber in the flight path of the ions in which a suitable electrical field is generated by applying voltages to plates inside the chamber. The voltages are kept electronically at a constant ratio to the accelerating voltage (E) during the determination of normal spectra. The field of the electrostatic analyser acts as an energy monochromato allowing, via a small slit, only ions with a much narrower range of kinetic energy than would otherwise be the case, into the magnetic field. The detailed calculations have produced two general designs, with somewhat different properties. These are shown schematically in Figs. 4 and 5 and are known as the Nier–Johnson and Mattauch–Herzog geometries respectively.

The Nier–Johnson geometry results in the ions being focused onto a complicated three-dimensional surface. A collector is placed at the appropriate position on this surface and by varying (automatically) either the magnetic field (H) or the accelerating voltage (E), the ions of any one m/e value are brought onto the collector at one time and recorded electrically. Normally the magnetic field is allowed to decay exponentially when the maximum amplitude of the electrical signal for any one type of ion is proportional to the relative intensity of this ion type. A slit is placed in front of the collector to improve the resolution; the width of this slit can be varied to give greater or lower resolution (see p. 24).

The Mattauch–Herzog geometry results in the ions being focused onto a two-dimensior surface and hence, in this case, the easiest way to record the spectrum is to put a photographic plate in the focal plane, thus recording all the ions simultaneously. A collector working on the same principles as that described for the Nier–Johnson geometry can also be used.

Time-of-flight mass spectrometry

Since to a first approximation all the ions have the same energy, each type of ion will have the same velocity, different from any other type of ion. Therefore, if the ions are allowed to drift down an evacuated tube they will reach a collector placed at the end at different times. To avoid interference between ions produced at different times, the ions are produced in pulses either by switching on the electron beam or accelerating voltage for a very short time. This type of mass spectrometer is often used when very fast scan speeds are desired. The resolution of these instruments is not very high, being limited by the decay time of the collector system and the kinetic energy spread of the ions. The principle of operation of this instrument is shown schematically in Fig. 6. The pulse of ions is shown at different times (t_0-t_4); each dot represents an ion of a different mass.

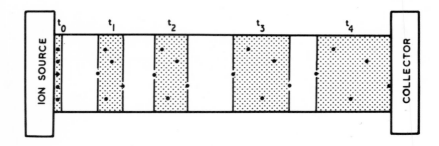

FIG. 6. Mode of operation of time-of-flight mass spectrometer.

Collecting and recording of ions

As can be seen from the above discussion, in the case of double focusing mass spectrometers the method of ion recording is somewhat dependent on machine design. In this section the advantages and disadvantages of the two main types of collector/recording systems used in this case will be summarized. A brief mention of the system for the time-of-flight case will be made.

Photoplate

This records the whole spectrum simultaneously and therefore relative ion intensities are not affected by sample supply fluctuations (see p. 3). However, the response of the photoplate is dependent on the mass of the ion[1] and this results in the signal recorded not being a true measure of the ion's intensity. Further, it cannot be used to study the effects of various factors such as electron voltage (see p. 28) on one particular ion. It requires a microdensitometer/travelling microscope to measure the intensity and distance apart of the signals on the photoplate. Even when this procedure is automated it is relatively time-consuming and it thus cannot be used for data processing in combination with a real-time computing system. This, coupled with the problem of storing enough plates inside the machine, could lead to problems during a multi-component GC run.

Collector plate

This gives out an electrical signal which can be amplified in several ways. The most sensitive arrangement is to use an electron multiplier as a collector. This, however, discriminates against higher m/e ions. The signal from the collector can be displayed in a variety of ways — on a photographic chart using galvanometers, a tape recorder, a pen recorder, an ammeter or an oscilloscope. The latter is used in the case of time-of-flight mass spectrometers and the 'scope signal' is photographed. The data can also be fed from the mass spectrometer directly to the computer for so-called real-time computer processing. Alternatively, the data can be stored on the tape recorder and fed into the computer when convenient. The collector also allows the study of one ion (see p. 28). The main disadvantage is that it does require a constant sample supply throughout the determination of the spectrum. This, unless high speed scanning techniques are available, can occasionally be a problem with direct insertion techniques. Even with high speed scanning there is still a problem with the combined GC/MS link-up.

In summary, the actual arrangement selected depends on the use to which the mass spectrometer is put and the money available.

2
Basic aspects of organic mass spectrometry

PROCESSING OF A MASS SPECTRUM

Whether a mass spectrum is recorded electrically or on a photoplate it will consist of signals of varying intensities separated by time or distance. Examples of a spectrum recorded electrically and on a photoplate are shown in Figs. 1 and 2 respectively. The electrically recorded spectrum has been transferred onto a photographic chart by means of galvanometers. Apart from storage on magnetic tape these two methods and the use of a pen recorder are those most commonly employed. In the latter case the spectrum looks very like one trace of the electrically recorded spectrum shown in Fig. 1. In the photoplate, which shows the peaks corresponding to the various $[M - Cl]^+$ ions, each set of lines represents exposures for different times for the same ion.

In the case of the photographic chart the recorder incorporates a number of galvanometers with pre-set relative sensitivities. The amplified signal from the ion collector is fed to all of these. The resulting traces on the chart increase in sensitivity from A to F. This makes it possible to measure the height of very small peaks in the least sensitive trace A fairly accurately. It is also necessary for the detection of the very low intensity peaks labelled 'metastable peak' in the specimen spectrum.

Whatever the means of recording the spectrum, the first step in processing the spectrum is the assignment of m/e values to each peak. This can be done in a variety of ways.

Assignment of m/e values

Use of an electronic mass marker
This can only be used when either the magnetic field (H) or the accelerating voltage (E) is varied (scanned) to produce the spectrum. In the former case a sensitive magnetic flux metre is used which, by suitable calibration and electronics, will 'print out' a measure of the m/e values onto the recording device employed. Direct measurement of the accelerating voltage (E) will, in a similar way, also allow the measurement of m/e values. The use of a crystal clock in conjunction with a reference compound will be discussed later (see p. 25).

By inspection
If the above method cannot be used, a starting point for m/e assignment by direct counting must be recognized. To help in this the spectrum is determined over a fixed mass range so that the m/e value of the lowest ion is known. With a photoplate, correct positioning will achieve this; in the case of magnetic or electrostatic scanning, starting or stopping at a

BASIC ASPECTS OF ORGANIC MASS SPECTROMETRY

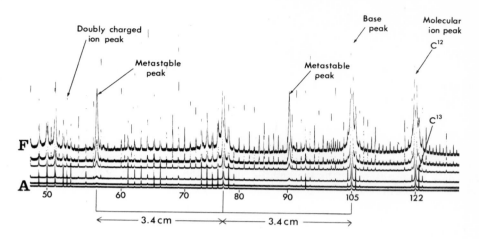

FIG. 1. Electrically recorded spectrum of acetophenone.

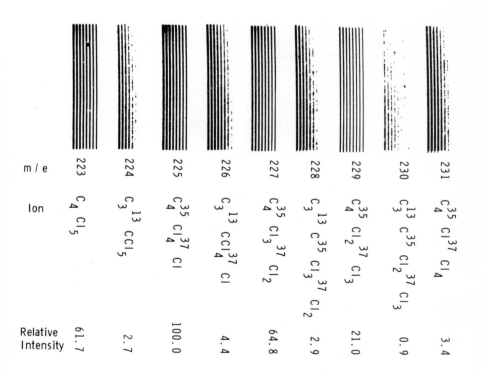

FIG. 2. $[M - Cl]^+$ region of hexachlorobutadiene spectrum recorded on a photoplate. (Reproduced by kind permission of Dr. Desiderio, Baylor College of Medicine, Houston, Texas).

fixed value of H or E, as appropriate, will give the same result. Systematic counting then allows the determination of the m/e values of the other ions.

There are certain other aids to the location of the starting point and counting. In the case of an exponentially decaying magnetic scan the separation between adjacent m/e values decreases with increasing m/e values. Thus, by comparing an unknown spectrum with one that has already been counted, the ions in the low mass region of the unknown spectrum can have their m/e values determined. The low mass region is used since this effect is most noticeable here. Another aid to starting and counting is the fact that there are normally no ions between m/e 18 (H_2O^+) and m/e 24 (C_2^+) nor between m/e 2 (H_2^+) and m/e 12 (C^+). The presence of a pair of peaks in the low mass region separated by four m/e units and with intensity ratios of 5 (low mass):1 (high mass) is often due to the presence of air i.e. m/e 28 (N_2^+) and m/e 32 (O_2^+) in the machine. This can often be used to identify a starting point. If there are gaps in the high m/e region it is necessary to use a linear extrapolation procedure using the distance between the peaks on either side of the gap.

Other ways

In the case of storage on a magnetic tape or direct computer (real-time) processing a known spectrum is often recorded simultaneously with the unknown spectrum. The computer is programmed so that it recognizes the ions in the high mass region due to the known spectrum. By an extrapolation procedure the m/e values of the other ions are then determined. The computer also checks its starting point and extrapolation by confirming the presence at the appropriate places of the ions expected from the known spectrum. A similar method can be used in the *Inspection method* described on p. 10. In both cases it is preferable that the known spectrum has ions at higher m/e values than those in the unknown spectrum.

Other points

In the above discussion it was assumed that all the ions had a unit charge and were formed before acceleration. In practice this is not always the case. Peaks corresponding to doubly-charged ions are often found. If these ions have an even mass then the m/e value will be an integer and their presence will not affect the assignment of m/e values. However, if they have an odd mass they will occur at $(n + 1)/2$ m/e units where n is an integer. This can normally be recognized by an abrupt change in the peak spacing. Similar remarks apply to triply-charged ions, but these are less often encountered. These doubly- and triply-charged ions are often associated with poly-unsaturated or aromatic molecules (see p. 19).

When an ion is formed after acceleration, it will not be focused according to Eqn. (1) on p. 5. These fragmentations give rise to the so called metastable peaks in the spectrum (see Fig. 1). These peaks are usually broader than the normal peaks and occur mostly at non-integer m/e values. Because of their width they normally cause no trouble in the m/e assignment process. Their use will be described later (see p. 23).

Data processing and presentation

This involves measuring the intensity of each peak and presenting this and its m/e value together in some appropriate form. In the case of the metastable peaks their m/e value is quoted to ±0.1 m/e units; if the intensity is required special procedures have to be adopted. The measurement of intensity has been discussed in Chapter 1 on p. 9. In the case of electrical recording the peak area is the measure of the relative ion abundance, assuming no discrimination in the mass spectrometer or amplifying systems. With an exponentially decaying magnetic scan the peak widths are all the same and hence the peak height is proportional to the ion intensity.

After the intensities of all the peaks have been measured it is sometimes necessary to remove the contributions due to machine background i.e. compounds already present in the machine. To do this a spectrum should be determined under the same operating conditions just before the sample vapour is allowed into the ionization region. Subtraction of the intensities of the various peaks in this spectrum then gives the correct intensities. The spectrum (i.e. intensities) is then normalized; conventionally the largest peak in the spectrum is assigned the value of 100 per cent. An alternative method of presentation is to express the intensity of one ion as a percentage of the total ion current intensity ($\%\Sigma_m$) where m is the mass of the lowest ion considered. The total ion current is merely obtained by adding together, before or after normalization, the intensities of all the ions down to mass m in the spectrum.

This data processing can be done using a computer, either by feeding the data directly to it, or via a magnetic tape. The data after determination of m/e values and intensities can also be fed to the computer.

Normally these data are stored and presented in one of two ways, either in tabular or line diagram form.

m/e	% Abundance	m/e	% Abundance	m/e	% Abundance
25	4·1	51	10·1	76	0·3
26	7·2	52	7·2	77	6·1
27	25·1	53	2·1	78	100·0 (M$^{\ddot{+}}$)

FIG. 3. Tabular form of data presentation.

The tabular form
In the tabular form shown in Fig. 3 the ion intensities are expressed as percentage relative abundances; a similar table can be used if the intensities are expressed as $\%\Sigma_m$. This has the advantage of giving the exact abundance of an ion on inspection. The disadvantages are that the salient features cannot be seen at a glance, and that the presentation space required is relatively large.

The line diagram
The line diagram shown in Fig. 4 allows the ion intensities to be expressed in both forms simultaneously. It is space saving and shows the overall features at a glance. The important peaks usually have the m/e value written on as shown. In comparing a series of related compounds this method is particularly useful.

In either case the molecular ion is often labelled M$^{\ddot{+}}$, the $^{\ddot{+}}$ symbolism showing that the ion is positively charged and contains one unpaired electron. In some cases the term 'percentage relative intensity' will be found instead of 'percentage relative abundance'; the two terms are used inter-changeably in mass spectrometry.

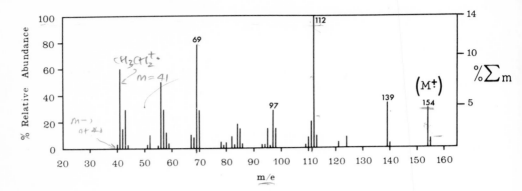

FIG. 4. Line diagram form of data presentation.

IMPORTANT SPECTRAL FEATURES
Isotopes

The isotopes of an element differ in the number of neutrons in their nucleus. Thus ions containing different isotopes of the same element have different m/e values. The relative intensities of the peaks in the mass spectrum at these m/e values are a measure of the abundance ratio of the isotopes in the ion. Unless the molecule has been deliberately enriched in one isotope this ratio will correspond to the natural abundance ratio. Thus the mass spectrum of a compound containing one chlorine atom will have two molecular ion peaks with an abundance ratio of 3:1 corresponding to $R^{35}Cl^{\ddagger}$ and $R^{37}Cl^{\ddagger}$. The chemical atomic weight of chlorine (35.5) can be seen to correspond to a mixture of these two isotopes in this ratio. Fragment ions containing one chlorine atom give the same result. The relative natural abundance of the isotopes of some common elements, their m/e values and the appearance of the grouping of isotopes in the spectrum of the element are shown in line diagram form in Fig. 5.

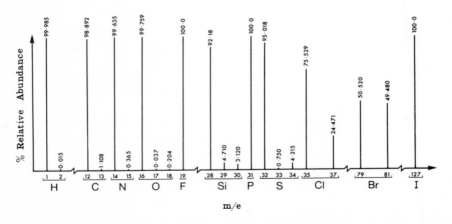

FIG. 5. Relative natural abundance of the isotopes of some common elements in line diagram form.

The exact natural abundance ratios for the isotopes of all elements can be found in most books of physical constants.[1]

It can be seen that when a molecule contains a carbon atom there is a 1.108 per cent chance of it being a ^{13}C and a 98.892 per cent chance of it being ^{12}C. Therefore, a molecule like methane, for example, will have a molecular ion which corresponds to $^{12}CH_4$ and another corresponding to $^{13}CH_4$. Their relative abundances are approximately 99:1. Normally, when discussing the nature of ions formed from a compound in a mass spectrometer only the ion containing the most abundant isotopes of the elements is mentioned (^{12}C, ^{1}H, ^{16}O, ^{14}N, ^{32}S, ^{35}Cl etc.).

When more than one carbon atom is present in an ion the probability of any one of these atoms being ^{13}C is found by using a binomial expansion:

$$(a.^{12}C + b.^{13}C)^n$$

where a = percentage natural abundance of the ^{12}C isotope
b = percentage natural abundance of the ^{13}C isotope
n = the number of carbon atoms present in the ion

In the case of an ion containing two carbon atoms this expression becomes:

$$(98.89.^{12}C + 1.11.^{13}C)^2$$

since the percentage natural abundances for ^{12}C and ^{13}C are 98.89 and 1.11 respectively. Expansion of the above expression and subsequent normalization of the ion intensities gives:

$$9779.^{12}C^{12}C + 224.^{12}C^{13}C + 1.25.^{13}C^{13}C \equiv 100.(^{12}C)^2 + 2.20.^{12}C^{13}C + 0.01.(^{13}C)^2$$

In these calculations the terms in which the sum of the superscripts are the same are collected together i.e. in the first term the sum is 24. This ensures that all ions having the same unit m/e value, whether they contain the same isotopes or not, are collected together. The importance of this will become clear when the case of ions containing two elements both having isotopes is discussed. In the case above the first term is due to the ion containing only ^{12}C atoms and the ion will have an m/e value corresponding to 24, plus the sum of the weights of the other elements in the ion. In the second term the sum of the superscripts is 25 and the ion contains one ^{12}C and one ^{13}C atom. Since the sum of the superscripts is one more than in the first case this ion will occur at one m/e unit higher. If the first ion were a molecular ion, then the second ion would be known as the $[M + 1]^{+}$ ion. As can be seen the last term corresponding to a $[M + 2]^{+}$ ion is very small and can be ignored unless n is large. As a rough guide the peak height at $M + 1$ is 1.1 per cent of that at m/e M for each carbon atom present. For a sesquiterpene ($C_{15}H_{24}$), $n = 15$, and the height of peak at m/e $M + 1$ in its mass spectrum is approximately 16.5 per cent of the peak height at m/e M. If we consider a molecule with 100 carbon atoms, it can be seen that there is a higher probability of having a ^{13}C atom present than there is of not having one (intensity of $[M + 1]^{+} \approx 110$ per cent of M^{+}). The isotopes ^{15}N, ^{17}O and ^{2}H (D) also contribute to the peak height at m/e $M + 1$ when they are present. Their magnitude can be seen from the line diagram of isotope ratios (Fig. 5). The isotopes ^{18}O and ^{34}S contribute to the ion current at m/e $M + 2$ peak, the ^{34}S being more obvious because of its

4 per cent abundance. Allowing for all possible isotopic species in the peaks at m/e M, M + 1 and M + 2 etc. in the group of peaks corresponding to the molecular ion in the mass spectrum of a molecule, it is possible to calculate the atomic combination giving rise to the relative abundance ratios observed in this group. This calculation, however, is involved and is normally unnecessary when high resolution mass measurement is available. Values for percentage abundances for $[M + 1]^+$ and $[M + 2]^+$ ions for certain common cases are given in Ref. 2(a).

Atoms in which the abundant second isotopic species contains two more neutrons e.g. chlorine and bromine will give very characteristic intensity ratios at m/e values of m, $m + 2, m + 4$ etc. When a molecule contains any of these atoms, groups of peaks will be present in the mass spectrum corresponding to each ion containing one of the isotopic species. The abundance ratios of the group of peaks corresponding to the molecular ions M^+, $[M + 2]^+$, $[M + 4]^+$ etc. show the number of atoms of this type in the molecule. The abundance ratios of the groups of peaks at $m, m + 2, m + 4$ etc. corresponding to each fragment ion will indicate the number of atoms of this type in the fragment ion. Using a similar binomial expansion to that already discussed for the $^{12}C/^{13}C$ case the number and relative abundances of peaks can be calculated for any combination of atoms of this type present in an ion. Consider the binomial expansion of $(a.^fX + b.^gX)^n$ where n is the number of the atoms of the element present in the ion and ^fX and ^gX are two isotopes of mass f and g ($g = f + 2$). In the case where $n = 3$, we obtain on expansion and collection of terms in which the sum of the superscripts is the same:

$$a^3.(^fX)^3 + 3a^2b.(^fX)^2.(^gX) + 3ab^2.(^fX).(^gX)^2 + b^3.(^gX)^3$$

There are four terms in this expansion and in general they will be $n + 1$ terms. If the case of chlorine is considered, then $^fX = {}^{35}Cl$, $^gX = {}^{37}Cl$, $a:b = 3:1$ and the above expansion becomes:

$$27.(^{35}Cl)^3 + 27.(^{35}Cl)^2.(^{37}Cl) + 9.(^{35}Cl).(^{37}Cl)^2 + (^{37}Cl)^3$$

Thus, for any ionic structure containing three chlorine atoms there will be four peaks in the abundance ratios 27:27:9:1 and since the difference between the sum of the superscripts in those ions of nearest mass is two, the peaks will be separated by two m/e units.

Similarly for the case of bromine where $^fX = {}^{79}Br$ and $^gX = {}^{81}Br$, the ratio for $a:b$ can be taken as 1 and when $n = 3$ we obtain:

$$(^{79}Br)^3 + 3.(^{79}Br)^2.(^{81}Br) + 3.(^{79}Br).(^{81}Br)^2 + (^{81}Br)^3$$

Thus, by analogy with the chlorine case, any ionic structure containing three bromine atoms will give rise to four peaks having abundance ratios of 1:3:3:1 and separated by two m/e units.

When an ionic structure has two different elements each with a pair of isotopes, then the expression for the number and relative abundances of the peaks is as below, where the nomenclature is obvious by analogy with the preceding expression.

$$(a.^fX + b.^gX)^n.(c.^hY + d.^kY)^q$$

Various possibilities can occur, but the following example shows the general method.
Consider the case of an ionic structure containing two chlorine atoms and one bromine atom. The above expression then becomes:

$$(3.^{35}Cl + {}^{37}Cl)^2.({}^{79}Br + {}^{81}Br)$$

which on expansion gives:

$$9.({}^{35}Cl)^2.({}^{79}Br) + 6.({}^{35}Cl).({}^{37}Cl).({}^{79}Br) + ({}^{79}Br).({}^{37}Cl)^2$$
$$9.({}^{35}Cl)^2.({}^{81}Br) + 6.({}^{35}Cl).({}^{37}Cl).({}^{81}Br) + ({}^{81}Br).({}^{37}Cl)^2$$

By collecting terms where the sum of the superscripts is the same, the following expression is obtained in which m denotes the weight of the ion containing two ^{35}Cl and one ^{79}Br atoms:

$$9.(m) + (6 + 9).(m + 2) + (1 + 6).(m + 4) + (m + 6)$$

Thus four peaks are seen in the spectrum separated by two m/e units and having relative abundance ratios of 9:15:7:1. The abundance ratios for many of the common cases are given by Beynon.[1] A few examples of various combinations of the two elements bromine and chlorine are given in Fig. 6.

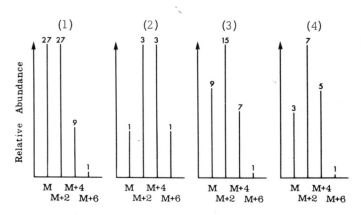

FIG. 6. Intensity ratios of halogen combinations: (1) 3 chlorine, (2) 3 bromine, (3) 2 chlorine and 1 bromine, (4) 1 chlorine and 2 bromine.

Computer programs can often be profitably employed in cases where many isotopes are present e.g. organometallic compounds, since many metals have several isotopes. In such cases hand calculation is tedious and prone to error.

The molecular ion $M^{\ddot{+}}$

The ion formed by loss of one electron from a molecule is known as the *molecular ion*. The corresponding peak in the mass spectrum is called the *molecular ion peak*. The first information the mass spectrum will give about the compound under examination is its molecular weight and to obtain this the peak corresponding to the molecular ion must be recognized.

In the section dealing with isotopes it was implied that the molecular ion peak in the mass spectrum would be considered as the one corresponding to the molecular ion containing the most abundant isotopes. This peak is accompanied by peaks one and two mass units heavier due to molecular ions containing heavier isotope atoms. If the compound is pure and gives rise to a molecular ion peak this peak must be a member of the group of peaks of highest m/e in the spectrum. A useful rule is enunciated in the introduction to the section on fragmentation i.e. 'the peak corresponding to the molecular ion of a molecule will be of even mass except when an odd number of nitrogen atoms is present'. The corollary of this statement is useful in recognizing molecules containing an odd number of nitrogen atoms.

Common impurities in compounds from natural sources are homologues which differ in mass from the molecular ion by $(n \times 14)$ where n is 1, 2, 3, etc. These are easily recognized because the difference of 14 mass units cannot represent a stable neutral fragment. When the spectrum shows peaks in the molecular ion region separated by a mass difference of between 3 and 15, either a mixture is present or neither represents the molecular ion, e.g. two peaks differing by 3 mass units could be produced by loss of a methyl group (15) and a molecule of water (18) from the molecular ion of an alcohol which is too easily fragmented to yield an observable molecular ion peak. There are many instances when a molecular ion peak is not recorded or is of very low abundance; instead a peak corresponding to the m/e value M + 1 is sometimes observed. This occurs in the spectra of some esters, ethers, amines, aminoesters and nitriles and is due to the abstraction of a proton from a neutral molecule by the molecular ion (an ion—molecule reaction). This results in the oxonium or ammonium type ion being formed and enhances the stability of the system. In cases where no molecular ion peak or a $[M + 1]^+$ ion is observed, analysis of the mass spectrum can lead to the mass of the molecular ion being deduced e.g. in the case of the alcohol cited, where two groups of peaks differing by three mass units were observed. The fact that fission can occur in some molecular ions so that a hydrogen is lost must also be borne in mind because this gives rise to a peak at m/e M − 1 in the molecular ion group of peaks. An aldehyde is an obvious example; other possibilities become clear when the basic factors governing fragmentation are appreciated. This illustrates the fact that at least partial analysis of the mass spectrum is always necessary to ensure that the molecular ion peak chosen is compatible with all the other features of the mass spectrum.

The percentage abundance of the molecular ion depends on its stability to decomposition and therefore is indicative of the type of molecule under examination. Conjugated olefines and aromatic systems often give rise to abundant molecular ions and this is often true in cases of other structures which can also be imagined to delocalize the positive charge. Cyclic molecules often have somewhat more stable molecular ions than their open chain analogues owing to the fact that more than one bond must be broken before fragmentation can occur. There are so many different types of compound and so many competing factors involved that it is not useful to list compounds in order according to the abundance of their molecular ion peaks. It is, however, useful to bear in mind that compounds having the molecular ion as the most intense peak (the base peak) in their spectra usually contain an aromatic system with no easily breakable bonds in the side chains or saturated rings attached to them. Compounds whose spectra have no molecular ion must have some special property to which this behaviour can be attributed e.g. very high stability of the fragment ion, the presence of a very labile bond or the stereochemistry of the molecular ion being such that a very facile rearrangement allows the breaking of an otherwise strong bond.

Doubly- and triply-charged ions

These have already been mentioned on p. 12. The presence of such ions gives some indication that the original molecule has delocalized electron orbitals. Thus aromatic molecules and those containing conjugated systems commonly give rise to doubly-charged ions. If the unit mass (m) is even for the ion, then its m/e value will be integer and the only way it can be identified in the high electron energy mass spectrum (approximately 70 eV) is by finding the doubly-charged $[M + 1]^{2+}$ species at half a m/e unit higher. If m is odd, then it occurs at a non-integral m/e value. Similar rules can be drawn up for triply-charged ions. In both cases relatively high electron energies are required to produce these ions and by reducing the electron beam energy to below 20 eV these ions will normally disappear.

Multiply-charged ions, in addition to indicating something about the properties of the molecule under investigation, are useful in revealing the m/e values of their singly-charged counterparts when these cannot be determined by direct counting.

Ions formed by inter-molecular processes

The percentage of molecules ionized in the vapour of a compound in the ionization chamber of a mass spectrometer is very low. Consequently there is a probability of a collision between a molecular ion and a neutral molecule during which an atom or group may be abstracted from the neutral molecule by the molecular ion with the formation of an ion which is heavier in mass than the molecular ion. The abundance of these ions depends upon the square of the sample pressure because they are formed by a second order process. Normally, at the low sample pressures used in routine work, these ions are of negligible abundance. The exceptions are cases involving hydrogen abstraction in which the ions formed will give rise to a peak at m/e M + 1. These ions are produced in significant amounts in cases where the molecular ion has low stability and fragments readily whilst the corresponding protonated molecular ion is very stable. This results in a very small or negligible molecular ion peak M^{+}, but a substantial peak at m/e M + 1.

Behaviour of this type is observed in the spectra of some ethers, esters, amines, aminoesters and nitriles. The $[M + 1]^{+}$ ion is useful for finding the molecular weight in cases of this kind. It is a disadvantage if one wishes to find the atomic constitution of a molecular ion using the isotope method applied to the peaks at m/e M, M + 1 and M + 2. Adiponitrile is an example of a molecule which exhibits an abundant $[M + 1]^{+}$ ion in its mass spectrum. There are two nitrile groups so that the probability of hydrogen abstraction by the molecular ion from a neutral molecule is increased.

Metastable peaks

The occurrence of this type of peak has been mentioned on p. 12 and the first spectrum on p. 11 shows several examples. The reason for the nature and position of these peaks lies in the focusing properties of the mass spectrometer. In the derivation of the focusing equation on p. 5 it was assumed that all ions have the same energy before entering the magnetic analyser. This is, however, only true for ions which are formed before they are accelerated (see p. 5). If an ion is formed by fragmentation after acceleration some of the kinetic energy of the original ion will be transferred to the neutral fragment which is expelled on fragmentation. Thus an ion of mass m_2 so formed will have less energy than an ion of the same mass which is formed in the source before acceleration. Calculations show that the maximum value of the intensity of the metastable peak will occur at a m/e value m^* given by:

$$m^* = (m_2)^2/m_1$$

where m_2 is the mass of the ion formed and m_1 the mass of the ion fragmenting. The process is described as a *metastable transition*, since the ions of mass m_1 involved are intermediate in stability between those ions of mass m_1 which fragment in the source and those which do not fragment at all. The ion m_1 which fragments is known as the *parent ion* and the ion m_2 is the *daughter ion*. It should be noted that these ions will only be detected if they are formed in the field-free region between source and magnet, but in the case of a double focusing mass spectrometer (see p. 7) metastable transitions occurring before the electrostatic analyser will not be recorded under normal conditions. This is because the electrostatic analyser is designed to reject ions not having the correct energy (see p. 8).

In comparison with reactions in the source metastable reactions are relatively slow, this being a characteristic of rearrangement processes. Intense metastable ions often correspond to rearrangement reactions in which small stable molecules such as water (18), carbon monoxide (28), ethylene (28) and propylene (42) are expelled. The figures in brackets indicate the mass of the neutral molecule expelled. Metastable peaks are also seen for processes involving the expulsion of a radical.

Determination of m_1 and m_2

As there is only one equation and two unknowns (m_1, m_2) various possible pairs of m_1, m_2 values must be tried in the equation to see if they give a value in agreement with the observed value of m^*. Agreement within ±0.2 mass units is normally considered good enough. Two examples of the trial and error method are given below.

1. The metastable in the spectrum of acetophenone (see Fig. 1) is at m/e 56.5. This was suspected to be (for reasons see below) due to the ion of m/e 105 $[C_6H_5CO]^+$ expelling carbon monoxide to give m/e 77 $[C_6H_5]^+$.

$$m^* \text{ (calculated)} = (77)^2/(105) = 56.47$$

and thus the assignment was correct.

2. The metastable in the spectrum of 3-methyl-*trans*-2-pentene at m/e 30.0. This was suspected to be due to the expulsion of a methyl radical from m/e 56 $[C_4H_8]^+$ to give m/e 41 $[C_3H_5]^+$.

$$m^* \text{ (calculated)} = (41)^2/(56) = 30.0$$

thus confirming the assignment.

Without some aids to help in the guessing this assignment would become tedious and so various aids have been developed. Computer programs can easily be written which will give all the possible solutions of m_1 and m_2 for a given m^*. Beynon[1] has described a

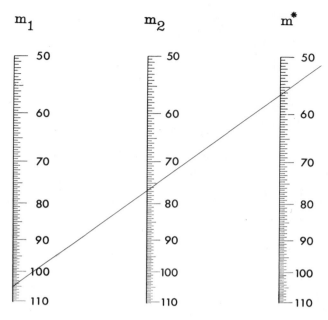

FIG. 7. The nomogram.

device called a nomogram which uses the fact that:

$$\log m_1 + \log m^* = 2 \times \log m_2$$

The nomogram consists of three equally spaced logarithmic scales corresponding to m_1, m_2 and m^* (see Fig. 7). A perspex ruler can be used to connect the observed value of m^* with possible values of m_2 and m_1. In Fig. 7 the diagonal line represents the ruler set, in this case, for the correct solution for the metastable at m/e 56.5 in the spectrum of acetophenone (see example 1). In the case where the origin of a particular ion is required, the ruler can be set on this value of m_2 and connected with various values of m^* observed to see if a possible m_1 value can be found. In this connection it should be noted that the mass of the fragment ion which would be lost in a proposed transition must correspond to a chemically possible entity i.e. a loss of between 3 and 14 mass units is unlikely. In the event that the formulae of the two ions are known (see p. 25), then the daughter ion must not contain more atoms of any element than the proposed parent ion. The first restriction can be easily written into the computer program mentioned earlier in this section.

Another method of guessing at the values of m_1 and m_2 is available in the case of a spectrum determined by using an exponentially decaying magnetic scan. Combining the scan law, the focusing law and the metastable equation it can be shown that, in this case, the distance on the chart between the parent and daughter ion is the same as that between the daughter and metastable ion. Thus a ruler can be used to check if this is so in the case of a potential m_1, m_2 pair. In the spectrum of acetophenone (see Fig. 1) this is indicated for the correct parent and daughter ions corresponding to the metastable peak at m/e 56.5, by the scale underneath.

For several reasons, but in particular to overcome the difficulty of having to guess m_1 and m_2 which can lead to several possible pairs for a metastable observed in the low mass region, a new technique for observing metastables has been developed. This can only be used in the case of a double focusing mass spectrometer, since it uses the focusing powers of the electrostatic analyser to observe metastable transitions in the region between the electrostatic analyser and the source. This region is known as the *first field-free region*. The principle used is to give the parent ion enough kinetic energy by increasing E, the accelerating field, so that when it fragments in the first field-free region the daughter ion so produced will have the correct kinetic energy to go through the electrostatic analyser. Using the magnet the daughter ion is tuned on to the collector and the magnetic field is left unchanged. The accelerating voltage is then increased until a new peak is detected on the collector. When the maximum signal intensity has been reached the ratio (R') of the accelerating voltage to the initial accelerating voltage is noted. Since H and R are constant (see p. 5) the m/e value of the ion corresponding to the parent ion is given by the simple relationship:

$$R' = m_1/m_2$$

Thus, in the case of the ion at m/e 57 in n-octane, peak maxima were detected for R' values as shown below, using this technique.

R'	m/e parent ion
1.492	85
1.737	99
1.983	113

The calculation for the first case is as follows:

$$1.492 = m_1/57$$
$$\therefore m_1 = 57 \times 1.492$$
$$= 85$$

When a collector is not fitted after the magnetic analyser a variation of this technique is used. In this the accelerating voltage is kept constant and the electrostatic analyser voltage scanned. A collector is fitted immediately behind the electrostatic analyser and a plot of signal intensity versus electrostatic analyser voltage obtained. The ratio (R'') of the final to the initial electrostatic voltage for an intensity maximum is given by:

$$R'' = m_1/m_2$$

However, since the m/e focusing (i.e. the magnet) is not used in this case, both m_1 and m_2 are unknown and, as in the case for metastables recorded in the normal spectrum, must be guessed.

Both these methods have the advantage that the normal spectrum is not seen. This enables higher amplification to be used to detect the weak metastable ion, without resulting in the overloading of the signal due to the ions formed in the source. This allows the detection of low intensity metastable ions which may have been obscured by the normal ions in the conventional mass spectrum.

Uses of metastable peaks

Ions related by a metastable peak indicate that the daughter ion (m_2) is formed from the parent ion (m_1). Thus the presence of an appropriate metastable gives confirmation of a proposed fragmentation route of the ion m_1 which may be very helpful. The absence of a metastable does not rule out a proposed scheme, since many reactions are completed in the source before acceleration. Since a metastable transition occurs in a small time interval (approximately 10^{-5} s) it is assumed that it represents a concerted elimination of one neutral species. However, a consecutive process occurring in the appropriate field-free region would give a metastable in the same position.

A second use of metastables is in the interpretation of the spectra of molecules which have been isotopically labelled to investigate fragmentation mechanisms. Often, since complete isotopic purity is difficult to achieve, the spectra are due to both the labelled and unlabelled molecule. A similar application is their use in examining the spectra of a mixture of two structurally different molecules. In both cases the metastables are used to sort out which ion comes from which molecular species. Relative intensities of metastable, parent and daughter ions can also be used to decide if ions of the same formula, found in the spectra of two different compounds, have the same structure and energy distribution. If the relative intensities are the same, then this is probably the case.

OTHER TECHNIQUES
High resolution mass spectrometry

So far only the unit m/e value of an ion has been determined. This could correspond to various ions, each containing different combinations of several elements, the sum of whose atomic weights would give the m/e value observed. A knowledge of which particular combination of elements was correct (i.e. the ion's formula) would be of considerable help in elucidating the fragmentation pattern of any molecule and, in the case of an unknown molecule, would give its molecular formula and perhaps, from its fragmentation pattern, its structure.

This can be accomplished as follows. Although for analytical purposes chemists normally only need to use molecular weights* correct to two decimal places e.g. on the carbon = 12.000 ...... scale the weight of oxygen normally used is 15.99 and of hydrogen 1.01, the isotopic weights are known to a much greater accuracy than this. A few examples are given below:

$$H = 1.007825 \qquad N = 14.003074 \qquad F = 18.998405$$
$$C = 12.000000 \qquad O = 15.994915 \qquad S = 31.972074$$

Taking m/e 28 as an example and considering only the above elements there are four combinations which could give rise to this unit m/e value:

$$CO = 27.994914 \qquad CH_2N = 28.018723$$
$$N_2 = 28.006158 \qquad C_2H_4 = 28.031299$$

Pairs of ions like N_2^+ and CO^+ which occur at the same unit m/e value in a spectrum are

---

*Corresponding to the naturally occurring mixture of isotopes.

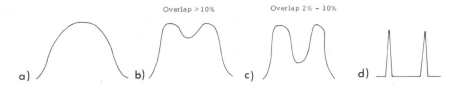

FIG. 8. Various degrees of resolution: (a) unresolved, (b) partial, (c) adequate, (d) high.

called a *doublet*. Groups of more than two such ions are known as *multiplets*. It can be seen from the table that if the m/e value for the ion(s) at m/e 28 could be measured accurately enough, then the formula(e) of the ion(s) present could be determined. In order to be able to measure the mass (= m/e for this purpose since e is a constant) of one of the ion(s), it is necessary to be able to separate the signal of this ion from the signals of the other ion(s) in the multiplet. In practice it turns out that two peaks of equal intensity are satisfactory provided that the two signals overlap at a point where the intensity is less than 10 per cent of the maximum intensity of one of the ions. This is known as the *x per cent valley resolution*. Examples of various degrees of resolution are shown in Fig 8. The quantity quoted is $m/\Delta m$, where Δm is the difference in the masses of the two ions for which a x per cent valley can be obtained. For a double focusing mass spectrometer a typical value for a 10 per cent valley would be 30,000. In the case of m/e 28 the value of Δm for the two ions nearest in mass (CO^+ and N_2^+) is 0.0112 and thus the resolving power, if a 10 per cent valley could be just achieved, would be:

$$m/\Delta m = 28/0.0112$$
$$= 2,500$$

This is not difficult to achieve with a good single focusing instrument. However, if there were two ions at m/e 280 of formulae $C_{18}H_{36}N_2$ and $C_{18}H_{36}CO$, then Δm would still be 0.0112, but the resolving power if a 10 per cent valley was obtained would be:

$$m/\Delta m = 280/0.0112$$
$$= 25,000$$

This sort of resolving power can be only achieved with a double focusing instrument and it is essential to measure the weights of both ions accurately. For a molecular ion which is a singlet a good single focusing machine will provide sufficient resolution to determine its weight accurately. Attempted mass measurement of such a doublet at m/e 280 with a single focusing instrument would result in an answer between the two values which would lead to confusion.

Factors involved in limiting resolving power

The upper limit of the resolving power of a mass spectrometer is limited by several factors:
1. The variation in kinetic energy of the ions which is reduced by the introduction of the electrostatic analyser.
2. The widths of the slits at the exit of the source and the entrance to the collector. In the higher resolution instruments these widths are normally variable and can be used

to control the resolution of the instrument. This is important since higher resolution reduces sensitivity and discriminates against metastable peaks.
3. The overall dimensions of the machine.
4. The divergence of the ion beam due to space charge effects.

Mass measurement

This is carried out by comparison with an ion of known mass produced from a reference compound. The masses of the two ions should be within 10 per cent of each other and preferably less. Perfluoro compounds are commonly used as reference compounds because they produce abundant ions over a wide mass range, whose masses are known accurately. Also fluorine is a mass-deficient element i.e. its accurate mass (18.9984) is less than its nearest integral mass (19), so that ions containing many fluorine atoms are always well separated from ions composed of the common elements like carbon, hydrogen, nitrogen and oxygen. Thus at m/e 69 the ion CF_3^+ (68.9952) is well separated from the commonly occurring ions at this m/e value e.g. $C_4H_5O^+$ (69.0263), $C_5H_9^+$ (69.0705). This makes identification of the ions due to the reference (marker) compound easy, which is important.

In one of the methods used the ratio of the two masses is determined; in the other it is the difference between the masses of the two ions which is calculated. Hence, knowing one mass, the other can be calculated simply. The first method depends on the presence of a collector and is rather slower but less expensive than the other.

The two methods are as follows:

Computerized determination of the accurate masses of all the prominent ions in the spectrum. The spectrum of the reference compound and that of the unknown are recorded simultaneously on the same recorder. If a collector is used, a time signal from a very accurate crystal clock is also recorded simultaneously on the same recorder. The difference in time between the recording of the maximum intensity of the two ions is a measure of Δm, the difference in mass between the two ion. When the spectra are recorded on a photoplate then the distance between the maximum intensity of the two ions is a measure of Δm. This distance is measured by means of an accurate travelling microscope/densitometer.

These data i.e. signal intensity versus time are then fed into a computer. In the case of a collector the data can be fed in as they are produced, so called real-time computer processing, or they can be stored first on a magnetic tape and fed in when convenient. Real-time computing is not possible with a photoplate.

In either case the computer is instructed how to recognize the ions of the reference compound and, from the difference in time or distance between the reference ion and the unknown ion, to calculate the mass of the latter. It then calculates the formula of the ion from the atomic weights of the elements present and prints out this information which can be presented in a variety of forms, the most common being the *element map*; a small portion of one is shown in Table 1.

In the left-hand column the computer prints out the unit m/e value of the peak. Each of the other columns represents one possible combination of heteroatoms which could be present in the original molecule and hence in the ion. The actual combination is shown by the column heading. When the computer finds a peak it calculates the mass of the ion(s) present. It prints out the m/e value of the peak and prints, in the column for the particular combination of heteroatoms found, the number of carbon and hydrogen atoms in the ion. The figure after the hyphen gives the difference between the observed mass and the calcul-

TABLE 1. Part of an element map of an alkaloid[a]

Unit m/e	CH	CHO	CHN	CHNO	CHN$_2$	CHN$_2$O
182			13/12-0 ++++	11/20-0 ++++++		
183					12/11-2 ++	
197					13/13-0 ++++	
213					14/17-0 +++	
269						17/21-1 +
326						21/30-0 +

[a]Reproduced from an article by K. Biemann, *Pure Appl. Chem.*, **9**, 95 (1964) by kind permission of the author and IUPAC.

ated one for the formula quoted. This error is in milli-mass units (10^{-3} g). The number of crosses gives a measure, on a logarithmic scale, of the intensity of the ion.

The main problem with this type of presentation is that if there are several heteroatoms present or large numbers of more than one heteroatom, then the element map becomes rather large. Thus, in the case of n atoms of one heteroatom and m atoms of another in one molecule, the total number of columns is given by:

$$(n + 1)(m + 1)$$

This, when $n = m = 6$, would give 49 columns.

Peak matching. This method requires a collector and is somewhat older and slower, since each ion has to be measured in turn. For a particular ion the reference compound's nearest ion is found by consulting its spectra. The lower mass ion of the two is then tuned onto the collector by varying the magnetic field (H). The accelerating voltage is then lowered until the other ion is focused onto the same spot on the collector. From the ratio (R') of the initial to the final accelerating voltage the mass of the unknown ion is calculated from the equation below, in which m_l is the mass of the low mass ion and m_h of the high mass ion:

$$R' = m_h/m_l$$

The formula of the ion is then determined by consulting computer-generated tables which give the masses of all the possible ions at that unit m/e value. In practice these tables are normally restricted in mass range and only consider ions containing carbon, hydrogen and a limited number of oxygen and nitrogen atoms.[2a,2b] Thus, with large masses or other elements calculation of the possibilities by hand or computer is necessary. As for the computer cases, it is essential that the presence of elements, other than those above, be known before ionic formula assignment is attempted. In some cases the presence of other elements may be recognized from their appropriate isotopic patterns (see p. 14).

The above description of these two methods is only a brief summary; the actual practical details vary from machine to machine. The two processes are outlined in a diagrammatic form on p. 27, in which an ion of unknown formula at m/e 184 is compared with the ion at m/e 181 ($C_4F_7^+$, mass 180.9882), the latter being in the spectrum of heptacosafluorotri-n-butylamine.

COMPUTER METHOD	PEAK MATCHING
Feed in intensity versus time or distance data. Computer recognizes m/e 181 as reference ion and m/e 184 as unknown	Select m/e 181 as appropriate reference ion for m/e 184, tune former onto collector and determine R' (see p. 26)
↓	↓
Computer determines $\Delta m = 3.0254$ which gives $m_h = 184.1936$	Using equation $R' = m_h/m_l$ gives $m_h = 184.1936$

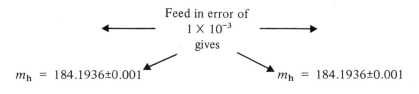

Feed in error of 1×10^{-3} gives

$m_h = 184.1936 \pm 0.001$ $m_h = 184.1936 \pm 0.001$

Knowing only C, H, N or O could be present

Computer calculates possible formula   Consultation of computer tables gives formula

which is $C_{11}H_{24}N_2$
mass = 184.1939

The conversion of an accurate ion weight into an unambiguous ion formula is restricted by two inter-related factors. The first is the accuracy of the mass measurement which at best is 1 part per million (p.p.m.) and more normally is about 5 p.p.m. In the best case this means that an observed mass of 200.1411 could correspond to an ion of mass between 200.1413 and 200.1409, whereas in the normal case the ion could have a mass between 200.1423 and 200.1401. This only gives an unambiguous answer if, with the elements which could be present, there is only one ion whose mass lies within these limits. The second related point is that the greater the number of elements or the greater the amount of one element present, the more likely it is that there will be more than one ion whose mass is inside the limits. Thus, with an error of 5 p.p.m. and only carbon and hydrogen present, ions of m/e greater than 500 can still have their formulae unambiguously assigned by mass measurement. If up to six nitrogen and six oxygen atoms could be present, then the m/e value for unambiguous assignment drops below 300.

✓ In the case of the molecular ion, the calculation of the intensities of the $[M + 1]^+$ isotope ions (given in Ref. 2(a)) corresponding to the natural abundance of ^{13}C etc. for each of the possibilities, and comparison with the observed value, often allows the selection of the correct ionic formula from the several possibilities given by mass measurement. This method cannot be used so easily with fragment ions since the peak one m/e unit higher may not contain only the isotope ions but other ions as well. If one possible formula for a fragment ion requires a larger isotope peak than observed, it can of course be ruled out. Sometimes the *Nitrogen Rule* (see p. 36) can be used, in the case of the molecular ion, to eliminate some of the possibilities.

Ionization potential

The ionization potential of a molecule is defined as the minimum energy required to remove an electron from the molecule to give the molecular ion. The electron impact ionization potential is defined as the minimum electron energy at which the molecular ion can be detected and the two values can differ slightly. When the energy transferred to the molecule by the electron impact exceeds the ionization potential by a sufficient amount, then fragment ions will be seen. The minimum electron energy at which a particular fragment ion $[F]^+$ is detectable is called the appearance potential $(A[F]^+)$ of the ion. The measurement of ionization and appearance potentials is a specialized topic in itself; for a detailed discussion see Ref. 3. These potentials are normally quoted in electron volts (1 eV = 23.05 kcal mol^{-1}) and are used in many branches of physical chemistry.

Ionization potentials are often useful in deciding where the charge is located in the ground state of the molecular ion, which in turn may help in the prediction of the fragmentation pattern of the molecule. Thus, since the ionization potentials of ethane, ethylamine and ethanol are 11.6, 9.5 and 10.7 eV respectively, it seems that in the ground state of the molecular ion of ethanolamine the charge is likely to be located on the nitrogen rather than elsewhere. From the above data it seems that, in general, charge location on nitrogen is preferred to charge location on oxygen, when the two heteroatoms are in similar environments. This is in keeping with the greater electronegativity of oxygen.

If we consider the fragmentation of the carbon—carbon bond in ethanolamine, then, since the charge is more easily located on the nitrogen, it would be expected that the

$$\overset{+\cdot}{NH_2}CH_2CH_2OH \longrightarrow \overset{+}{NH_2}:CH_2 \text{ or } CH_2:\overset{+}{OH}$$
$$\qquad\qquad\qquad\qquad\qquad m/e\ 30 \qquad\quad m/e\ 31$$

signal due to the nitrogen-containing ion at m/e 30 would be more intense than that due to the oxygen-containing ion at m/e 31. This is the observed result, m/e 30 being 10 times more intense than m/e 31. This is an example of <u>Stevenson's Rule which states that the charge will be preferentially located on the fragment ion of lower ionization potential.</u>

In general, the variety of problems which can be tackled using bond dissociation energies in conjunction with ionization and appearance potentials is considerable. Using these the ionization potential of the $[M-H]^+$ ion in the mass spectrum of toluene was calculated and compared with the value determined directly for the benzyl radical. The latter turned out to be lower. This stimulated the work which led to the discovery that the structure of the $[M-H]^+$ was not the benzyl cation, but the tropylium cation (see p. 47).

Isotope labelling

<u>Information about the structure of a molecule is obtained by comparing the mass of the molecular ion with the masses of the fragment ions formed from it by different decomposition paths. This yields potential formulae for all of these ions and for the neutral fragments lost during their formation.</u> The molecular ion is denoted $M^{\ddot{+}}$ and a fragment ion formed from it is denoted $[M-X]^{\ddot{+}}$, where X is the mass of the neutral fragment lost from $M^{\ddot{+}}$. Schematically, a few paths of fragmentation from the molecular ion can be represented as follows:

$$M^{\ddot{+}} \rightarrow [M-X_1]^+ + X_1$$
$$M^{\ddot{+}} \rightarrow [M-X_2]^+ + X_2$$

and so on. The fragment ions can also decompose further:

$$[M - X_1]^+ \rightarrow [M - X_1 - X_3]^+ + X_3$$
$$[M - X_2]^+ \rightarrow [M - X_2 - X_4]^+ + X_4$$

Fragmentation processes, and the factors which govern them are discussed later. It will suffice here to mention that a fragment ion of a particular mass can be formed by more than one process, or combination of processes. Mechanisms have been proposed to rationalize many fragmentation processes in terms of the movement of electrons, and of certain atoms in the decomposing ion. Specific bonds are broken and others are formed so that a stable ion and a stable neutral fragment are produced. Substitution of a heavier isotope atom in a particular position in a molecule for a normal atom will result in a gain in the mass of every fragment ion containing the isotope atom. Isotope labelling thus provides a means of confirming that a mechanism is operating because the fragments arising from the proposed mechanism and containing the isotope atom, can be predicted. Comparison of the mass spectrum of the normal compound with that of the labelled compound shows which peaks have shifted up the mass scale. The stable isotopes commonly incorporated into molecules as labels are ^{13}C, ^{18}O, ^{15}N and deuterium. The methods of introducing these into particular positions in different molecules are many and varied, and are beyond the scope of this book. Djerassi *et al.* give a good summary of these at the beginning of the second volume on structural elucidation by mass spectrometry.[4] Biemann,[5] who also covers the use of isotope labelling very thoroughly, gives the following very elegant example of deuterium labelling.

The elimination of an olefinic fragment from suitably substituted molecules like ethyl butanoate was proposed by McLafferty to occur as follows:

This would mean that a γ-hydrogen would be transferred to the oxygen atom by means of a six-membered transition state, and the neutral fragment atom eliminated contains the β- and γ-carbon atoms.

To prove this point three specifically labelled ethyl butyrates A, B and C were prepared and their mass spectra obtained. Note that the number of deuterium atoms incorporated is shown by the increment in mass of the molecular ion. The result of the McLafferty rearrangement in each case would be as shown over page.

The mass spectra in all cases showed the expected peak at the correct *m/e* value, thus proving the mechanism beyond all reasonable doubt. As shown above, the ions from A, B and C of *m/e* 90, 88 and 89 respectively are structurally suited for elimination of a further ethylene molecule by the same mechanism. Subtraction of 28 mass units (ethylene) in

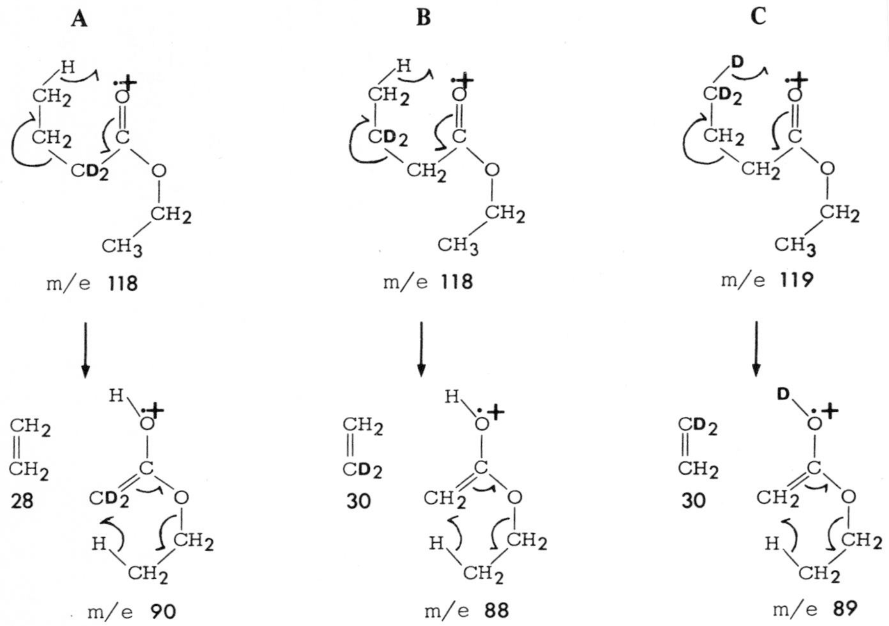

30 BASIC ASPECTS OF ORGANIC MASS SPECTROMETRY

[Structures A, B, C with m/e 118, 118, 119 respectively, fragmenting to m/e 90, 88, 89 respectively]

each case should yield peaks at m/e 62, 60 and 61 in the spectra of A, B and C respectively; these peaks are observed as expected.

Apart from solving mechanistic problems of the above type there are many uses of isotopic labelling combined with mass spectrometry in the field of organic chemistry. For example, to find the number of active hydrogen atoms in a molecule the mass spectrum of the unlabelled molecule is compared with the mass spectrum obtained after the active hydrogen atoms have been exchanged by deuterium atoms. In a ketone this is the number of hydrogen atoms α to the carbonyl group.

Nitrogen-15 labelling has been used in amino acid and peptide work with success. It is, of course, useful in settling many arguments involving a nitrogen-containing fragment. Carbon-13 is more difficult to introduce into a molecule but several skeletal problems involving hydrocarbons have been tackled by this method. Oxygen-18 is used when oxygen-containing functional groups are involved in a fragmentation process.

Quantitative determination of extent of isotope incorporation
When specific but incomplete deuteration or non-specific deuteration is carried out in a molecule a mixture will be produced containing a range of species deuterated to different extents. The percentage of each component present in this mixture can be calculated. During this calculation it is assumed that the relative abundances of the peaks in the molecular ion group of each species is the same, i.e. of the peaks at m/e M, M + 1 and M + 2.

The molecular ion group of peaks from the mass spectrum of the undeuterated molecule in Fig. 9(a) shows the relative abundances of the peaks at m/e M, M + 1 and M + 2 (not the percentage abundances). Figure 9(b) shows the corresponding group of peaks in the spectrum of the molecule after deuteration. In Fig. 9(b) the total intensity at the

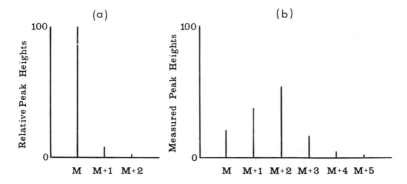

FIG. 9. Relative abundances of peaks from the spectrum of (a) an undeuterated molecule, and (b) the molecule after deuteration.

various m/e values is made up by contributions from each species as follows:

Undeuterated	M	M + 1	M + 2
Mono-deuterated	M + 1	M + 2	M + 3
Di-deuterated	M + 2	M + 3	M + 4
Tri-deuterated	M + 3	M + 4	M + 5

The percentage composition of the mixture of deuterated compounds whose mass spectrum is displayed in Fig. 9(b) is calculated as follows.

In the undeuterated species (Fig. 9(a)) the observed relative intensities at m/e M, M + 1 and M + 2 are 1:0.098:0.001. The observed peak heights for the deuterated mixture were:

m/e	M	M + 1	M + 2	M + 3	M + 4	M + 5
Intensity	20	32	43	14	1.00	0.01

The intensity at m/e M is due only to the undeuterated (d_0) species. Its contribution to the intensities at one and two m/e units higher are 20 × 0.098 = 1.96 and 20 × 0.001 = 0.02. The intensity of the mono-deuterated ^{12}C species is then 32 − 1.96 = 30 (corrected to two significant figures since this is the experimental accuracy) i.e. the intensity of the $^{13}C\,d_0$ species at m/e M + 1 has been subtracted from the total intensity at this m/e value. The d_1 species contributes 30 × 0.098 = 2.94 to the intensity at m/e M + 2 and 30 × 0.001 = 0.03 to the intensity at m/e M + 3. When the contributions to the intensity at m/e M + 2 from the d_0 and d_1 species are subtracted from the observed intensity, i.e. 43 − 2.94 − 0.02, this gives the intensity of the d_2 species as 40. Similarly the intensity of the d_3 species can be calculated to be 10. The contribution of this species to the intensity at m/e M + 5 is 0.01, which being the same as the observed value shows that there are no more highly deuterated species present. The corrected intensity and percentage of each component present in the mixture is shown in Table 2. The calculation of the latter in the case of the d_0 compound is 20 × 100/20 + 30 + 40 + 10 = 20. The peaks in Fig. 9(b) are therefore made up as shown in Fig. 10.

When the formula of the molecule being deuterated is known, the contributions to the

TABLE 2. The intensity of the ^{12}C deuterated species and the percentage of each component in the mixture.

Compound	Intensity	Percentage in mixture
d_0	20	20
d_1	30	30
d_2	40	40
d_3	10	10

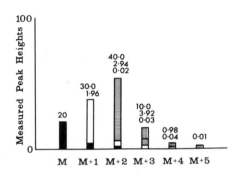

FIG. 10. Isotopic composition of peaks shown in Fig. 9(b).

intensity at m/e M + 1 from M due to the ^{13}C isotope is $n \times 1.11$ per cent, where n is the number of carbon atoms. The contribution to M + 2 is $n(n-1)(1.11 \times 10^{-2})^2 \times 100/2 \times$ M per cent. When nitrogen atoms are present 0.36 per cent of the intensity of M is added to the intensity of M + 1 for each one. Similarly, 0.20 per cent of the intensity of M is added to the intensity of M + 2 for each oxygen atom present. The abundance of naturally occurring deuterium can be ignored. When a fragment ion occurs at m/e M − 1 its isotope contributions to the intensity at m/e M and M + 1 must be taken into account, or this ion must be eliminated by the use of low electron voltages. Similar calculations can be carried out for ^{13}C and ^{15}N labelling, and with appropriate modification, for ^{18}O labelling.

Large numbers of experiments involving isotopically labelled molecules have been carried out in mass spectrometry; this section only indicates the type of way in which it is used. Biemann[5] has devoted a chapter to a thorough coverage of the subject, and this should be consulted if isotope labelling experiments are being contemplated.

Chemical pretreatment
Careful consideration of the problem in question before mass spectrometric examination can lead to a more suitable derivative being prepared to increase the amount of information obtainable.

For example, chemical pretreatment is useful to increase the volatility of a molecule, or to prepare a derivative with a characteristic fragmentation mode. The reactions used to prepare these derivatives should be applicable to very small quantities of material in many cases (approximately 1 mg). The economics of these reactions can be ignored. When the derivative has been selected its method of preparation is a problem for the chemist involved, who should be aware of all the latest techniques and the reagents available.

To increase the volatility
As explained earlier in the section on sample handling, a direct insertion probe is available on most modern instruments and enables the spectra of a wide range of organic molecules to be obtained. If, however, only a heated inlet system is available or the molecule has too low a vapour pressure to be examined, even using the direct insertion technique, it is essential to prepare a derivative. The melting point, compound type, and the known or expected mass of a molecule, are good rough guides to its volatility and hence suitability for examination. The mass spectroscopist in a particular laboratory will advise prospective users about potential samples in borderline cases. Some derivatives which have proved to be useful are shown in Table 3.

TABLE 3. Suitable derivatives for mass spectral analysis.

Compound	Derivatives
ROH	$ROCOCH_3$
ROH	$ROCH_3$
ROH	$ROSi(CH_3)_3$
RCO_2H	RCO_2CH_3
$RCONH_2$	RCH_2NH_2
RSO_3H	RSO_3CH_3

To make a derivative which has a characteristic mode of fragmentation
In some cases where the functional group does not have a sufficiently strong influence on the course of the fragmentation a derivative can be useful. An excellent example is the ethylene ketal derivative of a carbonyl group in a steroid molecule.[6] It can, of course, be used to advantage on most cyclic compounds containing a carbonyl group. The ethylene ketal directed fragmentation has been investigated by deuterium labelling which has enabled each of the many steps involved to be rationalized. The resulting ion contains the resonance stabilized system shown:

This gives rise to a base peak of the mass spectrum in which the other peaks are of low abundance. In some circumstances, the conjugated system of this ion can be extended or alkyl groups substituted on the double bond depending on the environment in the cyclic system from which the ion is derived. In some steroids several bonds are broken during

the formation of this type of ion. *N,N*-dimethylamine derivatives and ethylene thioketals give rise to exactly the same behaviour. The fragmentation involved in the formation of this type of ion is discussed in detail in the section on complex fission processes.

Since double bonds migrate under electron impact (see p. 45), the position of a double bond in a compound cannot be deduced from its mass spectrum. Several chemical pretreatment methods have been used to locate the double bond from the mass spectrum of a suitable derivative, of which the best seems to be as follows. The double bond is converted into the *cis* diol by treatment with osmium tetroxide. This, with acetone and acid, gives the isopropylidene derivative. The reaction mixture is then fed into the mass spectrometer

$$RCH=CHR' \xrightarrow{OsO_4} RCH(OH)CH(OH)R' \xrightarrow{(CH_3)_2CO}{H^+} \underset{\underset{C(CH_3)_2}{\overset{OO}{\diagdown\diagup}}}{RCH-CHR'}$$

$$\xrightarrow{e} [M-CH_3]^+, [M-R]^+ \;\&\; [M-R']^+$$

by means of a GC link-up. This allows the use of very small quantities (approximately 1 mg) of the olefine. From the mass spectrum which shows ions corresponding to $[M-CH_3]^+$, $[M-R]^+$ and $[M-R']^+$ the molecular weight can be determined from the m/e value of the first ion. Knowing this, R and R' can be determined from the m/e values of the $[M-R]^+$ and $[M-R']^+$ ions.

Oxidation and reduction can be useful in many cases e.g. a cyclic alcohol can be oxidized to a ketone and the latter converted to the ethylene ketal. The reduction of amides to amines has been used.

In some cases the molecular ion is non-existent in the spectrum (see p. 18), but by converting the compound to an appropriate derivative, it is often possible to observe its molecular ion and hence deduce the original formula of the molecule. Thus, in the case of long chain or polyhydric alcohols, the use of the trimethylsilyl derivatives (see p. 33) often allows the determination of the molecular formula. In general it is wise to check, if possible, the suspected molecular ion of a compound by converting the compound into an appropriate derivative and confirming that the molecular ion of the latter is in agreement with the molecular ion of the former and the chemical transformations executed.

These examples indicate that forethought can be useful when the mass spectrometric examination of a compound is contemplated. A knowledge of the factors which govern the fragmentation of molecular ions in a mass spectrometer is an obvious advantage in deciding if a derivative would be more suitable for examination than the compound itself.

3
Fragmentation of positive ions

INTRODUCTION

The further decomposition of the positive ions formed in a mass spectrometer has been studied for roughly the past twenty years, and during this time the mass spectra of thousands of compounds have been recorded. Use has been made of isotope labelling (especially with deuterium), metastable transitions, and of high resolution mass spectrometry. From this work it has been possible to gain some understanding of the factors which govern the breakdown of the positive ions produced in a mass spectrometer. These are the relative labilities of the bonds in the excited ion, the relative stabilities of the potential fragment ions and neutral fragments formed by competing fragmentation processes, and the probability of the ion breaking down by a rearrangement process involving concerted bond shifts through a cyclic transition state. It must be emphasized that the processes which will be discussed will only correspond to the most abundant ions in any mass spectrum because so many decomposition paths exist for even a small molecule.

There is a similarity between the fragments produced by the dissociation of the positive ions formed in a mass spectrometer, and those formed by other high energy processes like photolysis, pyrolysis and electrolysis. The lability of bonds and stability of positive ions and neutral fragments is often governed by the same factors with which the organic chemist is familiar from ionic organic chemistry. Thus any previous knowledge which the reader has about the above topics will be useful in the following discussion on fragmentation as well as in rationalizing the modes of fragmentation of a new class of compounds. It should, however, be remembered that some electron impact fragmentation processes are high energy reactions which will have no equivalent in ordinary ground state chemistry.

The main characteristic of molecules hit by high energy electrons (approximately 70 eV) is the initial formation of molecular ions with a high degree of vibrational and probably electronic excitation. In the absence of a more facile process any fragmentation reaction for which the steric factors are correct and the activation energy is less than 100 kcal mol^{-1} can give rise to a reasonably intense fragment ion peak. In some cases the molecular ion is very stable, as shown by an intense peak corresponding to the molecular ion. This can often be related to the fact that several bonds have to be broken before the expulsion of a stable neutral species can occur. Thus the molecular ion of naphthalene is very stable whereas the molecular ion of *t*-butyl chloride cannot be detected under normal conditions, the base peak in the spectrum being due to the *t*-butyl cation.

t-butyl cation $\quad CH_3-\overset{CH_3}{\underset{+}{C}}-CH_3 \qquad$ [naphthalene structure]$^+\quad$ naphthalene

In the case of fragment ions, if they are formed with sufficient excess energy, they will fragment further. Thus the ion at m/e 99 in the mass spectra of some ethylene ketals

m/e 99

is very stable whereas the $[M - CH_3]^+$ ion in the mass spectra of the isopropylidene derivatives used to locate double bonds, fragments further when the molecule contains an ester grouping. In the first case several bonds are broken in forming the ion which means that

the ion is probably formed with less excess energy than in the second case where only one bond is broken.

Normally a molecule is an even-electron species in which the electron spins are paired, whilst a molecular ion is an odd-electron species with one unpaired electron, one electron being removed in the ionization process. Since molecules are more stable than the corresponding molecular ion, it is deduced that even-electron all-paired species in general are more stable than odd-electron species. Hence, as observed, other things being equal, even-electron ions would be expected to be more prominent in mass spectra. Odd-electron ions are prominent when they have features leading to unusual stability e.g. naphthalene or when a particularly facile i.e. low energy process such as a McLafferty rearrangement (see p. 50) is responsible for their formation. Ions containing an even number or no nitrogen atoms are even-electron species when they occur at odd m/e values; if they occur at even m/e values they are odd-electron species. The reverse is the case if the ion contains an odd number of nitrogen atoms. This rule, the *Nitrogen Rule,* is due to the fact that the most abundant stable isotope of most elements with an odd valency has an odd mass, whereas the reverse is true of those elements with an even valency. The common exception to this rule among the non-metals is nitrogen which has an odd valency, but its most stable isotope has an even mass.

Many decomposition processes are thought to be initiated by the location of the charge/radical site on a particular atom or functional group in the molecule. This site will often be a functional group containing the heteroatom with the lowest ionization potential. One of the favoured fragmentation processes is the pairing of the electron on this site with an electron released by the homolytic cleavage of a bond attached to the carbon or other atom next to the group carrying the charge. This type of cleavage is known as an α-cleavage. The corresponding α-heterolytic cleavage results in the transfer of the charge. An example of each is shown below:

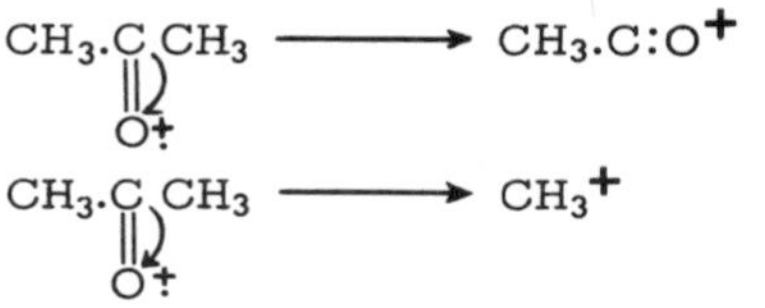

$$CH_3\text{-}\underset{\underset{\overset{+}{O}}{\|}}{C}\text{-}CH_3 \longrightarrow CH_3\text{-}C\text{:}O^+ \qquad \text{Homolytic cleavage}$$

$$CH_3\text{-}\underset{\underset{\overset{+}{O}}{\|}}{C}\text{-}CH_3 \longrightarrow CH_3^+ \qquad \text{Heterolytic cleavage}$$

The driving force for the first reaction is thought to be the energy gained by pairing the electrons in the ion which reduces the activation energy for the reaction. The acetyl cation formed in this reaction can be regarded as a resonance hybrid and other similar ions which can be formed in a similar way are shown. These types of ions are known as onium ions.

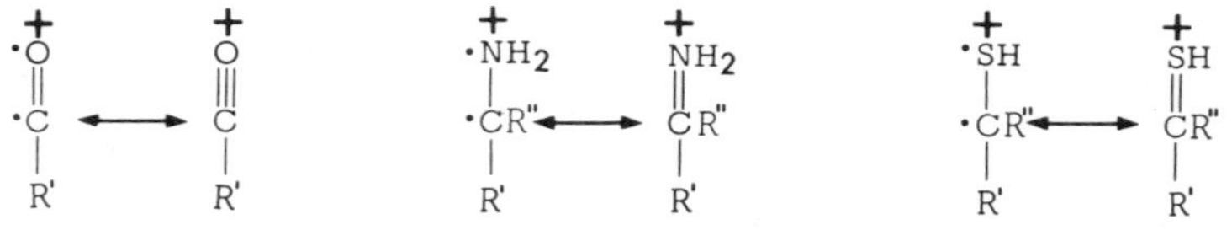

    oxonium   ammonium   sulphonium

Onium species can also be formed by cyclization reactions involving the expulsion of a radical from an atom more remote from the charge/radical centre. Thus in the case of the *ortho*-substituted phenylthioureas the loss of the substituent, which does not occur in the *meta*- or *para*-compounds, almost certainly occurs with the formation of the five-membered ring shown below:

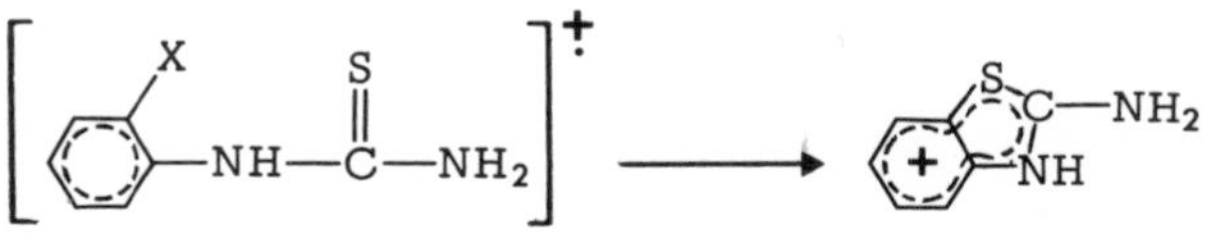

Other possible ring ions are shown below:

For saturated molecules containing the commonly occurring heteroatoms (N, S, O, Cl, I, Br etc.) the most favourable position for the charge/radical site is a lone-pair orbital of the heteroatom. The ionization potential for a lone-pair electron in acetone is 9.8 eV, for a π electron 10.6 eV and for a σ electron of the carbonyl bond 11.5 eV. These data suggest that the loss of an electron from the lone-pair then corresponds to the ground state of the molecular ion. The ground state of the molecular ion is the most heavily populated, and it is thus considered to be the one which leads to, i.e. directs, the major fragmentation pathways. However, since the molecule is bombarded with high energy electrons (approximately 70 eV), other electronic states can be formed and result in fragmentation reactions.

If the positive charge centre is localized it is labelled ($\overset{+}{\cdot}$) on the group when the ion is odd-electron and ($^+$) when it is even-electron. If one electron is transferred during a bond fission or bond formation this is denoted by a single headed arrow ⇀. A double headed arrow ⇀ indicates the moving of two electrons. This is the convention used by Djerassi et al.[7] The term α-fission or α-cleavage refers to the breaking of a bond adjacent to a functionalized carbon atom. The terms β-fission, γ-fission etc. refer to fission of the bonds proceeding away from the functional group in order. Ionized double bonds can be depicted as:

$$R-\overset{\cdot}{C}H-\overset{+}{C}H-R' \quad \text{or} \quad [R-CH=CH-R']^{\overset{+}{\cdot}}$$

In a case where the position of the positive charge is in doubt, $R^{\overset{+}{\cdot}}$ is used for an odd-electron ion and R^+ for an even-electron ion.

For fragmentation to occur a bond or bonds must break in the decomposing ion. When there is a choice of bonds, consideration of their relative labilities can indicate which is most likely to break. The polarizing of a carbon–carbon bond due to a halogen substituent increases its tendency towards cleavage in the decreasing order F>Cl>Br>I. The carbon–heteroatom bond fission, on the other hand, is governed by the electron affinity of the halogen atom which makes the carbon–iodine bond the most readily cleaved. Bonds allylic to double bonds and conjugated double bonds and β to aromatic systems are labile and are readily cleaved.

The charge/radical site in the ground state of the molecular ion is often located on a heteroatom in a functional group (e.g. hydroxyl, carbonyl, thiol, amino, ether and thioether). This means that the presence of a functional group on a carbon will weaken the α-bond(s) which attach that carbon to the rest of the molecule. This is due to the tendency to pair the unpaired electron on the heteroatom with one of the electrons of one of the α-bonds (see p. 37).

The degree of lability of this α-bond will depend on the heteroatom and its environment. If there is more than one such bond, then the bond preferentially cleaved will depend on the relative stabilities of the possible ions and neutral species which would be formed.

In a cyclic molecule substituted by a functional group of this kind the parts played by bond lability and fragment stability are well illustrated by consideration of the three isomeric methyl cyclohexanols. The 1—2 bond or 1—6 bond can break to yield the ions m/e 57 and m/e 71 respectively in 2-methyl and 3-methyl cyclohexanol. In the 4-methyl isomer only m/e 57 is recorded. The details of the mechanism will be discussed later under *Complex fission processes*. In the 2-methyl derivative the 1—2 bond is more labile than the 1—6 bond because the methyl group makes C-2 more highly substituted. This results in a secondary radical site at C-2 when the 1—2 bond breaks, whereas the radical will be primary on C-6 when the 1—6 bond breaks. Though the ion m/e 71 is more stable because of the methyl substituent, the ion m/e 57 is more abundant in the spectrum since 1—2 bond fission is more favoured. In the 3-methyl derivative both the 1—2 and 1—6 bonds are equally labile. The ion m/e 71 is, therefore, more abundant than m/e 57 because it is more stable. In the 4-methyl derivative only m/e 57 can result from the mechanism of fragmentation which has been rationalized using isotope labelling. This example illustrates the interplay of effects which finally determines the most favourable fragmentation path. However, the same argument applied to the case of 2-methyl cyclohexanone, in which the ions at m/e 55 and 69 are formed by a similar process, would suggest that m/e 69 should be less intense than m/e 55, whereas the opposite is, in fact, the case. Thus small changes in structure can alter the net result of the interplay of all the forces concerned.

In aromatic systems of the type shown below the electron withdrawing or donating effect of the substituent A can be transmitted through the ring to affect the lability of the bond B. The Hammet constants (δ, δ^+) were found to be in good agreement with the

extent of fragmentation occurring at B or with the appearance potential of the $[RCO]^+$ cation B, for a large number of substituents A.

These examples suggest that in some cases ground state rules can be applied to the rationalization of mass spectral fragmentation processes, although the actual processes may be more complicated.

The relative stabilities of the potential fragment ions and the corresponding neutral fragments which would be formed by possible competing fragmentation processes are very important in deciding which pathways will be most favourable for the fragmentation of a positive ion. This is particularly so for simple cleavage reactions in which the transition state is very close to the products. In order to try and measure the relative ease of a fragmentation it should be remembered that the intensity of the corresponding ion may not directly reflect its ease of formation, since it may fragment further. To identify primary fragmentation processes of a molecular ion it is useful to determine the spectra at lower electron beam energies (10—20 eV) which will normally suppress further fragmentation.

There are two types of fragmentation processes which will be considered separately.

SIMPLE FISSION PROCESSES
In such processes, a neutral fragment is lost by the breaking of a bond.

Saturated hydrocarbons
It was originally imagined that the even-electron ions of formula $C_n H_{2n+1}$ found in the mass spectra of long chain hydrocarbons were formed by simple fission processes. In the

case of the low mass ions of this type this is now known not to be true, but the ions formed by the expulsion of a methyl group from the molecular ions of such compounds are formed by a simple cleavage process. It is found that, if there is a tertiary centre, i.e. $R_1 R_2 R_3 C\text{–Me}$ where $R_1, R_2, R_3 \neq H$, then the $[M - Me]^+$ ion is more intense than the $[M - Me]^+$ ion formed from the molecular ion of an isomer of the first compound, but which has only a secondary centre. The ion formed by the loss of a methyl from the straight chain isomer, i.e. containing only a primary centre, is even less intense. In this case the ratio $(M - Me)/(M)$ i.e. the intensity of the $[M - Me]^+$ ion divided by that of the molecular ion M^+ is taken as the true intensity of the fragment ion. This observation is in keeping with the well known order of stability of carbonium ions, tertiary>secondary >primary. This point is illustrated in the mass spectra of n-octane and 2-methylheptane which are shown in Figs. 1(a) and 1(b).

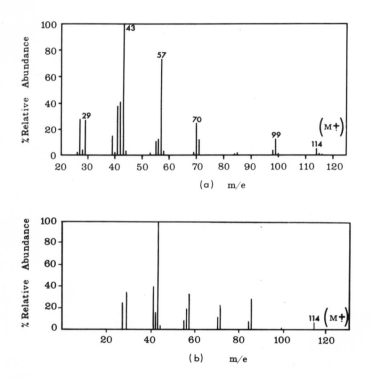

FIG. 1. Mass spectra of (a) 2-methylheptane and (b) n-octane.

Molecules in which the functional group is attached by a single bond from a heteroatom to a carbon atom

In such a molecule where the functional group may be NRR′, SR, OR, Cl, the α-cleavage reaction mentioned on p. 37 is normally a prominent primary fragmentation. This reaction is the homolysis of a bond which is attached to the carbon atom to which the functional group is also attached. The ground state in the molecular ion of these types of compounds is considered to be that in which an electron is missing from one of the lone-pair orbitals of the heteroatom. The activation energy for this reaction is thought to be relatively low due to the pairing of the remaining electron with that released by the homolysis. This is represented as shown below:

$$R-\overset{\overset{+\cdot}{X}}{\underset{R''}{C}}\cdot\cdot R' \longrightarrow R-\overset{\overset{+}{X}}{\underset{R''}{\overset{\|}{C}}} + \ \cdot R'$$

Examples of this reaction include the formation of the ion at m/e 30 in the mass spectra of primary amines, of m/e 31 in the mass spectra of primary alcohols and of m/e 45 in the mass spectra of n-alkyl methyl ethers; these ions are shown below:

$$CH_2=\overset{+}{N}H_2, \ CH_2=\overset{+}{O}H, \ CH_2=\overset{+}{O}Me$$

In this case R' = R" = H; if this is not so, the ions will appear at appropriately higher masses. Where there is more than one alkyl group which can be lost (i.e. R' and/or R" ≠ H) of the simple cleavage ions possible the one corresponding to the loss of the largest alkyl group is normally the most prominent. This is illustrated by the spectrum of secondary butanol. The effect may be due to the slightly greater stability of the larger alkyl radicals

$CH_3CH_2\overset{\cdot}{} \ + \ \overset{\overset{+}{OH}}{\underset{CH_3}{\overset{\|}{C}-H}}$	$CH_3CH_2-\overset{\overset{+}{OH}}{\underset{+\ CH_3\overset{\cdot}{}}{\overset{\|}{C}-H}}$	$CH_3CH_2-\overset{\overset{+}{OH}}{\underset{CH_3}{\overset{\|}{C}}} \ + \ H\overset{\cdot}{}$
m/e 45	m/e 59	m/e 73
% Abundance 100	19	1·2

with central precursor:

$$CH_3CH_2\overset{|}{\underset{\underset{CH_3}{|}}{+}}\overset{\overset{\cdot+}{OH}}{C}\overset{|}{+}H$$

lowering the activation energy of the reaction and thus making it a little faster. Alternatively it may be due to the greater ease of subsequent fragmentation of ions of this type which contain the larger radical group. These ions are known to fragment by a subsequent rearrangement reaction.

When different functional groups of this kind are present in the same molecule there will be competition between them to direct the fragmentation processes which are open to the molecule. The relative abundances of the peaks in the mass spectrum of a compound of this type, corresponding to the fragment ions formed by these competing processes, will depend on the relative stabilities of these fragment ions. As mentioned in the section on *Ionization potentials* the relative abundances of the peaks corresponding to $CH_2=\overset{+}{N}H_2$ and $CH_2=\overset{+}{O}H$ in the spectrum of ethanolamine are in the ratio of ten to one. Although the ionization potential arguments indicated that the charge in the ground state of the molecular ion was located on the nitrogen rather than the oxygen atom, the stability of the ions formed must also be considered. The stability of the unsubstituted ions of this type depends upon the heteroatom. The decreasing order of stability is $CH_2=\overset{+}{N}H_2 > CH_2=\overset{+}{S}H > CH_2=\overset{+}{O}H > CH_2=\overset{+}{F}$. When the hydrogen atoms of these ions are substituted by other groups, these can alter the stability. Electron-donating groups will increase the stability of the fragment ion while electron-withdrawing groups will decrease this stability. It is difficult to illustrate this point for electron-withdrawing groups (e.g. HCO, CH_3CO) because usually they introduce a competing fission process which masks the effect. However, methyl groups are good electron-donating groups and illustrate the effect of increasing the stability of the fragment ion in this way. In Table 1, in each case two ions, produced by retention of the charge on each heteroatom respectively, are shown. The two radicals also produced have been omitted for convenience.

TABLE 1. Effect of methyl groups on relative abundances of ions.

MOLECULE	ION I	ION II	ABUNDANCE RATIO I/II
$\begin{array}{cc} CH_2-CH_2 \\ \mid \quad \mid \\ OH \quad OH \end{array}$	$\begin{array}{c} CH_2 \\ \parallel \\ \overset{+}{O}H \end{array}$	$\begin{array}{c} CH_2 \\ \parallel \\ \overset{+}{O}H \end{array}$	1:1
$\begin{array}{ccc} CH_3-CH-CH_2 \\ \mid \quad\quad \mid \\ OH \quad\quad OH \end{array}$	$\begin{array}{c} CH_3-CH \\ \parallel \\ \overset{+}{O}H \end{array}$	$\begin{array}{c} CH_2 \\ \parallel \\ \overset{+}{O}H \end{array}$	9:1
$\begin{array}{cc} CH_2-CH_2 \\ \mid \quad\quad \mid \\ OCH_3 \quad OH \end{array}$	$\begin{array}{c} CH_2 \\ \parallel \\ \overset{+}{O}CH_3 \end{array}$	$\begin{array}{c} CH_2 \\ \parallel \\ \overset{+}{O}H \end{array}$	7:1
$\begin{array}{ccc} CH_3-CH-CH_2 \\ \mid \quad\quad \mid \\ OH \quad\quad OCH_3 \end{array}$	$\begin{array}{c} CH_3-CH \\ \parallel \\ \overset{+}{O}H \end{array}$	$\begin{array}{c} CH_2 \\ \parallel \\ \overset{+}{O}CH_3 \end{array}$	Both ions are of the same mass and isotope labelling would have to be used to find the ratio
$\begin{array}{cccc} CH_3-CH-CH-CH_3 \\ \mid \quad\quad\quad \mid \\ OCH_3 \quad OH \end{array}$	$\begin{array}{c} CH_3-CH \\ \parallel \\ \overset{+}{O}CH_3 \end{array}$	$\begin{array}{c} CH-CH_3 \\ \parallel \\ \overset{+}{O}H \end{array}$	6:1
$\begin{array}{c} \quad\quad CH_3 \\ \quad\quad \mid \\ CH_2-C-CH_3 \\ \mid \quad\quad \mid \\ OCH_3 \quad OH \end{array}$	$\begin{array}{c} CH_2 \\ \parallel \\ \overset{+}{O}CH_3 \end{array}$	$\begin{array}{c} CH_3 \\ \mid \\ C-CH_3 \\ \parallel \\ \overset{+}{O}H \end{array}$	1:5

With increasing substitution the competition of other fission processes must be kept in mind when anomalies occur in considering abundance ratios.

This type of argument applied to mixed substituted compounds like amino alcohols, aminomercaptans, hydroxymercaptans and their alkylated derivatives, shows the same type of variation, e.g.

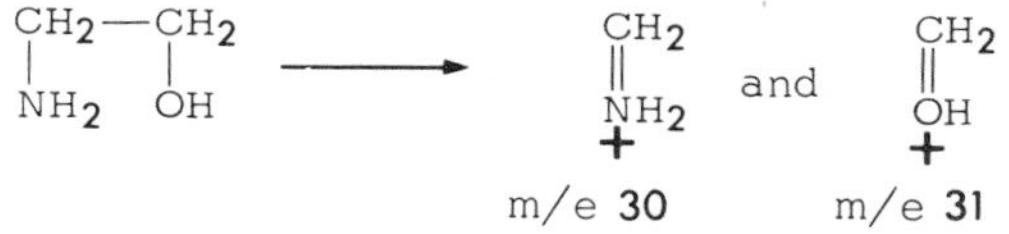

$$\begin{array}{c} CH_2-CH_2 \\ | \quad\quad | \\ NH_2 \quad OH \end{array} \longrightarrow \begin{array}{c} CH_2 \\ \| \\ NH_2 \\ + \\ m/e\ 30 \end{array} \text{ and } \begin{array}{c} CH_2 \\ \| \\ OH \\ + \\ m/e\ 31 \end{array}$$

Abundance ratio m/e 30:m/e 31 is 10:1

Substitution of two methyl groups on the amine carbon gives:

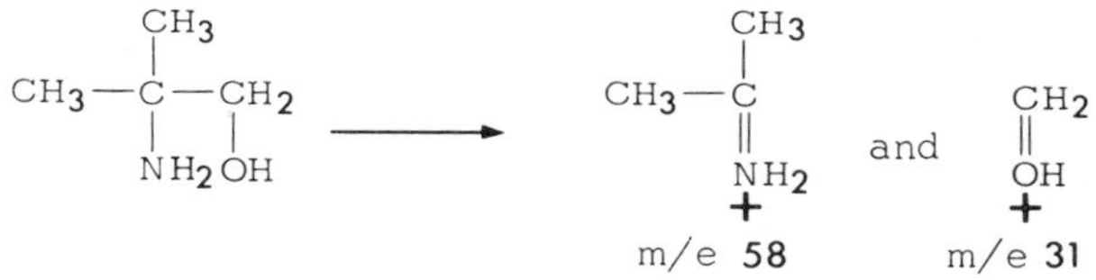

$$\begin{array}{c} CH_3 \\ | \\ CH_3-C-CH_2 \\ | \quad\quad | \\ NH_2\ OH \end{array} \longrightarrow \begin{array}{c} CH_3 \\ | \\ CH_3-C \\ \| \\ NH_2 \\ + \\ m/e\ 58 \end{array} \text{ and } \begin{array}{c} CH_2 \\ \| \\ OH \\ + \\ m/e\ 31 \end{array}$$

Abundance ratio m/e 58:m/e 31 is 20:1

The spectrum of the molecule substituted with two methyl groups on the hydroxyl carbon is not available but by the above arguments m/e 59 should be equal to, or greater than, m/e 30.

Peaks corresponding to this simple fission process are observed in the spectra of alcohols, ethers, amines, mercaptans, thioethers, and to a lesser extent, in the spectra of halogen-containing compounds, especially those containing fluorine and chlorine. If any of the functional groups in the compounds listed above are ring-substituted on a complex molecule, they will initiate the simple fission process described above as the first step of a complex fission process (see section beginning on p. 62).

Molecules in which the functional group is attached by a single bond from the heteroatom i.e. R—X. Fission of the carbon—heteroatom bond

Fission of the carbon—carbon bond can also occur and is represented as shown below where X can be F, Cl, Br, I, NR_2', SR' and OR'. If X is a halogen, process (1) will lead to an alkyl ion, while process (2) will lead to a halide ion. Process (2) is most favoured with iodides, iodine having the lowest ionization potential of the halogens. The structure of R will determine to what extent process (1) competes with process (2) because it determines the stability of the ion R^+.

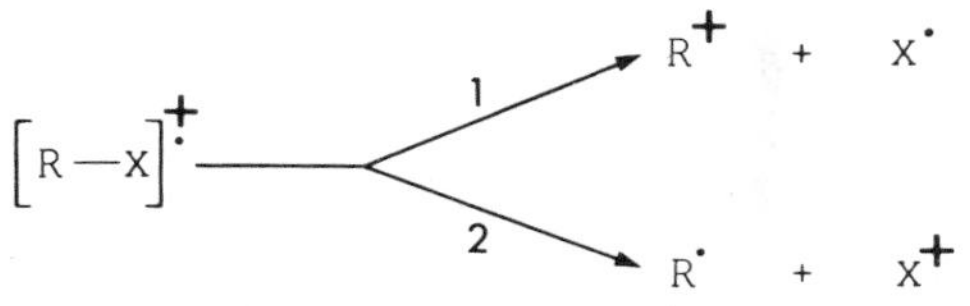

$$[R-X]^+ \begin{array}{c} \nearrow^{1} R^+ + X^\cdot \\ \searrow_{2} R^\cdot + X^+ \end{array}$$

The functional group X can be NR'R", SR' or OR', the last being the commonest and most important case. R' can be hydrogen only when R is an exceptionally stable group like *t*-butyl, because the neutral fragment OH˙ is not energetically favourable. In other cases, R' is methyl or greater. The processes (1) and (2) will lead to an alkyl and alkoxy ion respectively. Formation of the alkyl ion by process (1) is the favoured process because the alkyl ion is more stable than the alkoxy ion which contains the electronegative oxygen atom. The case of diisobutyl acetal illustrates normal behaviour with respect to C–X

$$CH_3-\overset{+}{\underset{\parallel}{C}}-H \quad \overset{CH_3}{\underset{CH_3}{\diagdown}}CH \quad \xleftarrow{B} \quad CH_3-\overset{A}{\underset{B}{\overset{\mid}{C}}}\!\!\!\begin{matrix}B & O+CH_2-CH \overset{CH_3}{\underset{CH_3}{\diagdown}} \\ & \\ B & O+CH_2-CH \overset{CH_3}{\underset{CH_3}{\diagdown}} \\ & A \end{matrix} \quad \xrightarrow{A} \quad \overset{+}{C}H_2-CH\overset{CH_3}{\underset{CH_3}{\diagdown}}$$

m/e 101   174   m/e 57

fission. The base peak of the spectrum is the ion m/e 57 formed by cleavage of one of the bonds labelled A whilst the corresponding ion m/e 117 is not significant. The ion m/e 101 formed by cleavage of bond B is the next most abundant in the mass spectrum, while the corresponding ion m/e 73 is insignificant.

Thioethers also participate in this type of fission, but process (2) is favoured. Owing to the greater capacity of sulphur to accommodate the positive charge, ions of the type RS⁺ are formed.

Molecules containing a functional group with a carbon–heteroatom double bond, i.e. R'–C(=X)–R'. Fission of the α-bond to the C=X group

With this type of functional group in which the heteroatom is commonly oxygen or sulphur a prominent simple fission process is that of the α-bond.

The unpaired electron is in one of the lone-pair orbitals of the heteroatom and this is paired by homolysis of the α-carbon bond (see p. 37) to give the ion. If more than one

$$R-\overset{\overset{+\cdot}{X}}{\underset{\parallel}{C}}\!\!\curvearrowright\!\!\cdot R' \quad \longrightarrow \quad R-\overset{\overset{+}{X}}{\underset{\parallel}{C}} \quad + \quad \cdot R'$$

group R can be lost then the ion corresponding to the loss of the larger radical is more prominent. This allows differentiation between aldehydes and ketones because in the spectrum of an aldehyde a peak will be present at m/e 29 corresponding to the ion HC≡O⁺. e.g. acetaldehyde

$$CH_3 \!+\!\! \overset{\overset{+\cdot}{O}}{\underset{\parallel}{C}}\!+\!\! H \quad \overset{1}{\underset{2}{\diagup\!\!\diagdown}} \quad \begin{matrix} CH_3\cdot \;+\; \overset{\overset{+}{O}}{\underset{\parallel}{C}}\!-\!H \\ m/e\ 29 \\ \\ CH_3-\overset{\overset{+}{O}}{\underset{\parallel}{C}} \;+\; \cdot H \\ m/e\ 43 \end{matrix}$$

The peak m/e 29 is the base peak in the mass spectrum of acetaldehyde because the formation of CHO^+ plus a methyl radical is more favourable energetically than the formation of CH_3CO^+ and a hydrogen radical. From the spectrum of a ketone the position of the carbonyl group can be found by consideration of the peaks $RC\equiv O^+$ and $R'C\equiv O^+$, since both will be present. For example, a peak at m/e 43, and a peak at m/e 71 would indicate that the ions $CH_3C\equiv O^+$ and $C_3H_7C\equiv O^+$ were present and, if the molecular ion of the spectrum was at m/e 86, that the compound was

$$CH_3-\underset{\underset{O}{\|}}{C}-C_3H_7$$

Confirmation of this, and information regarding the structure of the alkyl group are obtained by consideration of the rearrangement peaks (discussed later). If the peaks m/e 43 or m/e 71 are suspected to be $C_3H_7^+$ or $C_5H_{11}^+$ respectively, this point can be clearly proved by using exact mass measurement.

The spectra of esters in which one of the groups, say R', is alkoxyl show a peak corresponding to loss of OCH_3 or a homologue as well as loss of the alkyl group R from the molecular ion depending on the type of ester group present. In cyclic systems containing

$$C_2H_5\overset{1}{+}\underset{\underset{2}{\|}}{\overset{\overset{\cdot +}{O}}{C}}\overset{}{+}OCH_3$$
$$m/e\ 88$$

$$\begin{array}{c} 1 \nearrow \\ \\ 2 \searrow \end{array}$$

$$C_2H_5-\underset{\underset{O}{\|}}{\overset{+}{C}} \quad + \quad {}^\cdot OCH_3$$
$$m/e\ 57$$

$$C_2H_5{}^\cdot \quad + \quad \underset{\underset{O}{\|}}{\overset{+}{C}}-OCH_3$$
$$m/e\ 59$$

a $C=X$ group in one of the rings the α-fission process discussed above initiates the fission of a ring carbon–carbon bond. A complex fission process follows (see p. 62).

Molecules containing a carbon–carbon double bond
When a double bond is present in a molecular ion, fission of the carbon–carbon bond β to the double bond is favoured. The resulting carbonium ion is stabilized by resonance.

$$\left[C=C-C+R\right]^{\overset{+}{\cdot}} \longrightarrow C=C-\overset{+}{C} \longleftrightarrow \overset{+}{C}-C=C$$

By itself, however, this effect is not sufficient to ensure that the peaks in the mass spectrum corresponding to ions of this type will be characteristic. One reason is that double bonds appear to be able to migrate easily in the molecular ion thus rendering the spectra of different olefine isomers very similar. If, however, another structural feature is present in the molecular ion which blocks the migration path, stabilizes the olefine ion produced, or increases the lability of the bond being broken, the fragmentation becomes more characteristic. The molecular ion shown below illustrates these points.

$$[\text{structure}]^{+\cdot} \longrightarrow \begin{array}{l} [\text{structure}]^{+} \quad m/e\ 69 \\ \\ [\text{structure with OH}]^{+} \quad m/e\ 71 \end{array}$$

Bond (1) is labile because it is allylic to two double bonds and the ion m/e 69 formed by its cleavage has two methyl groups reinforcing its stability. Bond (2) is labile because of its allylic position and of the position of the hydroxyl group. The ion m/e 71 formed by its cleavage is a conjugated onium ion with additional stability due to the methyl substituent. m/e 69 is the base peak in the mass spectrum of this compound and m/e 71 is very abundant. The question of the relative stabilities of these two ions cannot be answered by consideration of their relative abundances because the hydroxyl group present in the original molecule can be removed as water in the heated inlet system by thermal dehydration and from the molecular ion by loss of water by means of a rearrangement process which is discussed later (see p. 53).

Another double bond in conjunction with the first extends the π-electron system, and thus enhances the lability effect on the β-bond, and the stability of the fragment ion formed.

Simple α-fission initiated by an aromatic system

In aromatic systems the charge is regarded as delocalized over the aromatic system. Thus the aromatic nucleus is regarded as the functional group and the cleavage shown below as an α-cleavage, where X is O, NH, S, etc.

$$[C_6H_5XY]^{+\cdot} \longrightarrow [C_6H_5X]^{+}$$

Examples of simple α-cleavage in the case of substituted benzenes include the following where R is an alkyl or aryl group:

$$[C_6H_5COR]^{+\cdot} \longrightarrow [C_6H_5CO]^{+}$$

$$[C_6H_5OR]^{+\cdot} \longrightarrow [C_6H_5O]^{+}$$

$$[C_6H_5ONO]^{+\cdot} \longrightarrow [C_6H_5O]^{+}$$

Similar reactions occur in poly-substituted benzenes and other aromatic systems. In the case of six-membered aromatics such as pyridine which contain nitrogen, this is also the case. Here the extent of the reaction is dependent on the position of the substituent. Thus the reaction shown for a 3-alkyl pyridine is more prominent in this isomer than the corresponding reaction in the other two isomers. This has been related to calculations which show in the ground state of the neutral pyridine molecule that the highest electron

$$\left[\underset{N}{\underset{|}{\bigcirc}}\text{-CH}_2\!\!+\!\!R\right]^{\ddagger} \longrightarrow \left[\underset{N}{\underset{|}{\bigcirc}}\text{-CH}_2\right]^{+} \longrightarrow \underset{+}{\bigcirc} + \text{HCN}$$

density is at the 3-position. This suggests that there will be the greatest location of charge in this position in the molecular ion, thus enhancing the cleavage in this position.

In all these cases the ions normally decompose further, often by the expulsion of acetylene or hydrogen cyanide, to give the $[C_5H_5]^+$ cation at m/e 65.

At first sight it would appear that the formation of m/e 91 in the case of the monoalkyl benzenes, as shown below, was just another simple cleavage reaction:

$$[C_6H_5CH_2R]^{\ddagger} \longrightarrow m/e\ 91$$

However the process is more complicated because the alkyl benzene molecular ion first ring-expands to give a substituted cycloheptatriene molecular ion before cleavage occurs to give rise to the tropylium ion:

$$\underset{\text{CH}_2R}{\bigcirc} \xrightarrow{e} \underset{H}{\overset{R}{\bigcirc}} \xrightarrow{-R} \underset{m/e\,91}{\bigcirc}{}^+$$

That this was not just a simple cleavage reaction was first indicated by some appearance potential measurements (see p. 28), but the nature of the process was shown by deuterium-labelling experiments. Labelling the various positions of toluene with deuterium showed that on electron impact all the hydrogens became equivalent before one was lost to give m/e 91. Further, metastable evidence showed that the ion at m/e 91 lost acetylene to give m/e 65. In the case of the deuterium-labelled compounds, the loss of acetylene, mono-deutero- and di-deutero-acetylene occurred in the relative intensity ratios which would be expected if all the hydrogens had become equivalent. This was explained on the basis of the ring expansion shown above coupled with rapid hydrogen randomization.

Recently the doubly labelled ^{13}C compound shown below has been made and its mass spectrum shows that the process is even more complicated. The $[M-H]^+$ ion in this com-

$$\underset{^{13}\diagdown\!\!\diagup^{13}}{\overset{\text{CH}_3}{\bigcirc}}$$

pound loses $^{12}C_2H_2$, $^{12}C^{13}CH_2$ and $^{13}C_2H_2$ in relative intensity ratios which require that ring expansion occurs with completely random scrambling of all the carbon atoms. At least three mechanisms could account for this result.

In this case the driving force for the reaction seems to be the formation of an even-electron aromatic ion, the tropylium ion. A useful rule for recognizing aromatic species is the *Hückel Rule* which states that a cyclic set of carbon atoms is aromatic if it has $4n+2$ π-electrons ($n=0$ or integer). The tropylium ion has 6 π-electrons and is thus aromatic and relatively stable. The case of $n = 0$ corresponds to the cyclopropenium ion at m/e 39. This ion is seen in these spectra also and the stability of the radical and cation may explain the formation of the $[C_3H_3]^+$ and $[M - C_3H_3]^+$ in the spectra of pyrrole, thiophene and furan. Substituted five-membered heterocycles of this type also undergo α-cleavage, the charge being delocalized over the aromatic ring:

In Chapter 4 most of the common processes of fragmentation undergone by substituted aromatic compounds are summarized for convenience. Several of the processes, discussed here and subsequently, can occur by utilizing an aromatic nucleus to fulfil some of the conditions necessary for their mechanisms to operate.

A fission process which occurs in a fragment ion to eliminate a neutral molecule

When simple fission occurs in a molecular ion to produce a saturated hydrocarbon ion $RCH_2CH_2^+$, or a carbonyl-containing ion $R-C=O^+$, a neutral molecule can be eliminated from this ion by cleavage of a single bond. The mechanism involving a two electron shift is shown below. In case (2) R could be an aromatic system. The driving force for this

1) $R-CH_2-CH_2^+ \longrightarrow R^+ + CH_2=CH_2$

2) $R-C\equiv O^+ \longrightarrow R^+ + CO$

appears to be the stability of the neutral molecule coupled with the stability of the ion produced. The extent to which process (1) occurs in saturated hydrocarbons appears to be small. As can be seen, the resulting ion R^+ could have been formed by simple cleavage in the first place. Metastable peaks in many cases, however, confirm that two steps are occurring. Process (2) is a common fission process when a carbonyl group is involved, and can account for abundant peaks in the mass spectrum corresponding to alkyl ions.

REARRANGEMENT PROCESSES

Here and elsewhere, since there is no all-embracing symbol for a rearrangement, the term (R) will be used. These processes involve bond formation as well as bond fragmentation before the expulsion of a neutral fragment and can be divided into two groups:
1. Migration of a radical e.g. hydrogen, phenyl etc. followed by fragmentation of one bond. A simple example is shown below:

$$[R(CH_2)_nCH_2OH]^+ \longrightarrow [M - H_2O]^+$$

This category includes those cases in which more than one of these radicals migrates.
2. Complex rearrangements in cyclic and other systems where fragments containing ring atoms, or atoms in the centre of a chain, are expelled, with or without group migration. This group includes reactions like the expulsion of carbon monoxide from the molecu-

lar ion of phenol or the loss of the central carbon atoms from a long chain hydrocarbon. Another example, shown below, is the expulsion of a methyl radical from tetralin:

$$\text{[tetralin]}^{+\cdot} \longrightarrow [\text{M}-\text{CH}_3]^+$$

The main factors influencing the nature and extent of rearrangement processes are as follows:

a) For a rearrangement to occur the stereochemistry of the parent ion must be such that the atoms and groups involved can come close enough together to interact. Studies on the McLafferty rearrangement[7] and the relative importance of rearrangement reactions in geometrical isomers have made this point clear. In any particular case the size of the transition state is dictated in part by this geometrical requirement and, for a hydrogen transfer, can vary from four to eight. This geometrical requirement means that the frequency factor in the rate constant for the reaction will be low compared with that for a simple cleavage. Hence for a rearrangement to compete with a simple cleavage it must have a lower activation energy. If the geometrical requirements are similar then the rearrangement with the lowest activation energy will predominate.

b) The following factors are important in determining the activation energy for a rearrangement. Firstly it is often low relative to simple cleavage reactions, since bonds are being simultaneously broken and formed. This can be particularly important in a case where the number of bonds broken is equal to the number formed. Another way of looking at this is to consider that a neutral molecule rather than a radical is being expelled. The McLafferty rearrangement and the retro-Diels-Alder reaction are good examples of this.

The difference in activation energy for two possible rearrangement reactions of one ion may be due to differences in steric strain in the two transition states. In other cases it may be the stability of the cation formed which is the important factor in favouring a rearrangement. The formation of the tropylium ion is one such case. In a rearrangement in which a neutral molecule is expelled, an even-electron ion will give rise to an even-electron ion and *vice versa* (see below). Since even-electron ions are preferred (see p. 36),

$$I^+ \xrightarrow{\text{Rearrangement}} I_1^+ + N$$
even electron   even electron   even electron

$$I^{+\cdot} \xrightarrow{\text{Rearrangement}} I_1^{+\cdot} + N$$
odd electron   odd electron   even electron

this may be why such an ion often fragments further by a rearrangement involving the expulsion of a neutral molecule.

These reactions can be conveniently classified and some of the more common types are discussed below.

Hydrogen rearrangements are often described as specific or random. A specific rearrangement is one which occurs by one mechanism, i.e. the hydrogen transferred comes from one particular site in the molecule. The McLafferty reaction is an example of this. In other cases more than one pathway exists, i.e. the hydrogen comes from more than one site.

To decide whether a reaction is random or specific, the molecule is labelled with deuterium in various positions and the resulting migration of the deuterium then shows whether the hydrogens came from more than one site. Some apparently random migrations may be due to the fact that the hydrogens are migrating (scrambling) in the parent ion before fragmentation occurs, thus obscuring whether the important hydrogen migration is specific or random. A fairly general rule seems to be that even-electron ions tend to undergo random hydrogen rearrangements whereas odd-electron ions tend to fragment via specific hydrogen migrations.

The McLafferty rearrangement

The McLafferty rearrangement is the most generally applicable specific rearrangement of all. The evidence for the proposed mechanism was discussed in the section on isotope labelling. The mechanism is as follows:

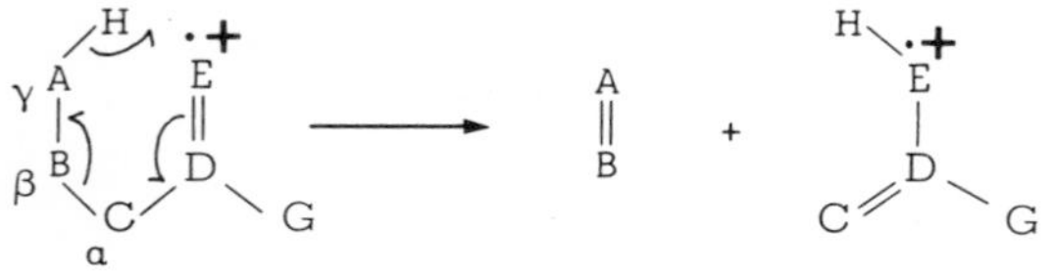

The essentials for this mechanism are a multiple bond DE and a γ-hydrogen available for transfer. The atoms A, B, C, D, E and the group G attached to D can vary widely. The spectrum of any molecule which fulfils the above conditions could exhibit peaks due to this type of rearrangement e.g. ketones, aldehydes, olefines, amides, nitriles, esters, substituted aromatic systems, phosphates, sulphites, etc. The examples shown below illustrate this mechanism:

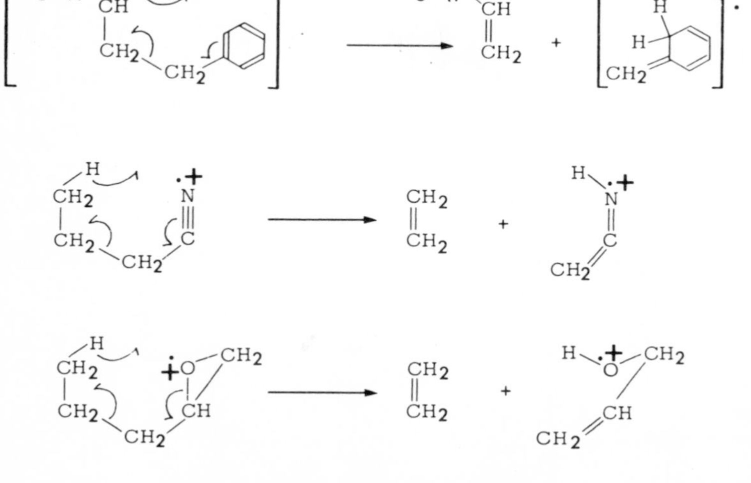

In the case of olefines and aromatic compounds the site from which the hydrogen is transferred cannot always be determined since extensive hydrogen scrambling occurs before fragmentation (see p. 47). This can sometimes be the case in other compounds as well. It is essential to label the hydrogen on A and show that this is the only hydrogen which moves before a reaction is called a McLafferty rearrangement. Labelling has shown, particularly in the case of even-electron ions, that what appeared to be a McLafferty rearrangement can be taken as evidence for the fragmentation of the B—C bond, but not for the presence of a hydrogen on A, the so-called γ-hydrogen.

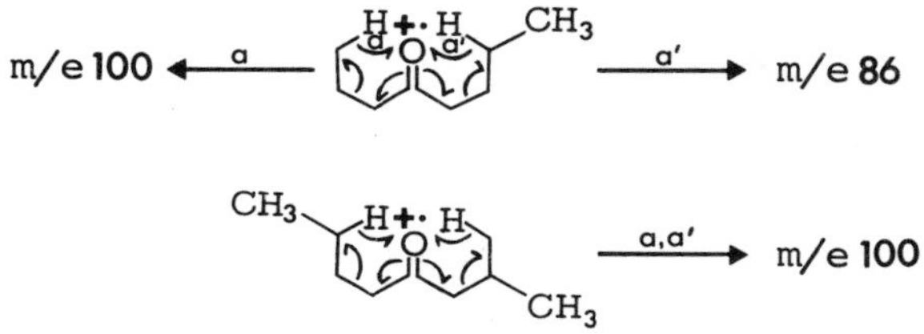

$$m/e\ 100 \xleftarrow{a} \quad \text{[structure]} \quad \xrightarrow{a'} m/e\ 86$$

$$\text{[structure]} \xrightarrow{a,a'} m/e\ 100$$

When a choice of pathways is available, the McLafferty ion corresponding to the elimination of the higher molecular weight olefine is more intense. Thus, in the case of n-propyl n-butyl ketone shown in the diagram, m/e 86 is more intense than m/e 100. This may reflect a greater ease of loss of a more stable, higher molecular weight olefine, or a slower subsequent rate of decomposition of the McLafferty ion containing the smaller radical. In the case of n-butyl isobutyl ketone, where a McLafferty rearrangement involving either radical would result in the elimination of the same olefine (propene), deuterium labelling evidence has shown that a secondary hydrogen is transferred ten times more easily than a primary hydrogen, the latter coming from the isopropyl methyl. This suggests that the key factor here is the slightly lower carbon—hydrogen bond energy for a secondary (*cf.* a primary) hydrogen.

Cases will be encountered in which a particular atom in the transition state or a substituent on one of these atoms may cause another process to occur so readily that this rearrangement is not favoured, even though the necessary structural conditions are fulfilled; this is simply the competition between various factors manifesting itself.

In esters greater than methyl the McLafferty rearrangement could occur involving the atoms of the alkoxy molecule. The structure of the molecule will determine if this is favourable and, if so, where the positive charge will reside after rearrangement e.g. ethyl benzoate eliminates ethylene, whereas n-butyl formate eliminates formic acid.

With butyl and higher esters the olefinic fragment can successfully compete for the charge because the ion formed from the olefinic part is of higher stability. *Stevenson's Rule* would predict that the olefinic fragment has an increased probability of getting the charge because alkyl or larger substituents will lower the ionization potential of the olefinic fragment. In terms of ionization potential the following examples show the trend:

$$[C_2H_5CO_2H]^{+\cdot}\quad 10\cdot5\ eV$$
$$[C_2H_4]^{+\cdot}\quad 10\cdot5\ eV$$
$$[C_4H_8]^{+\cdot}\quad 9\cdot7\ eV$$ 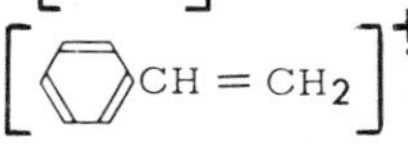
$$[C_6H_5CH=CH_2]^{+\cdot}\quad 8\cdot9\ eV$$

In β-phenyl ethyl esters the base peak corresponds to the olefinic fragment ion which, presumably, corresponds to a styrene molecular ion. The mechanism is exactly the same except that the charge remains on the olefinic portion and this is indicated by the use of the ⤴ arrow as shown below:

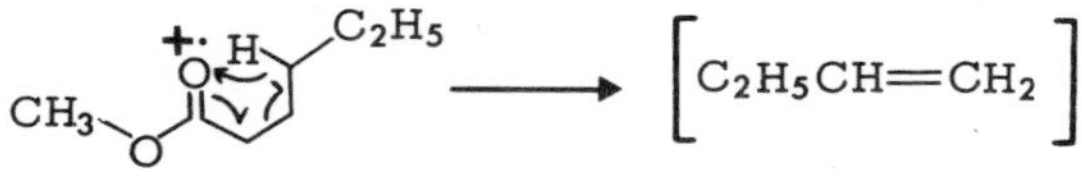

$$\text{CH}_3-\text{C(O)-O-CH(C}_2\text{H}_5\text{)-...} \longrightarrow [\text{C}_2\text{H}_5\text{CH}=\text{CH}_2]^{+\cdot} \quad m/e \ 56$$

The evidence so far suggests that a McLafferty rearrangement does not occur in an even-electron ion. Apparent cases of this have been shown not to be by using labelled compounds. A McLafferty rearrangement can occur in an odd-electron fragment ion containing the correct structural features. Thus, in the case of 2,6-diethylcyclohexanone, the odd-electron fragment ion at m/e 98 loses a second molecule of ethylene by a McLafferty rearrangement to give m/e 70. In 2,2-diethylcyclohexanone the second loss of ethylene does not occur. This suggests that the ion at m/e 98 in this case cannot de-enolize fast enough for a McLafferty rearrangement to occur, the enol form itself does not have the correct structure for a McLafferty rearrangement.

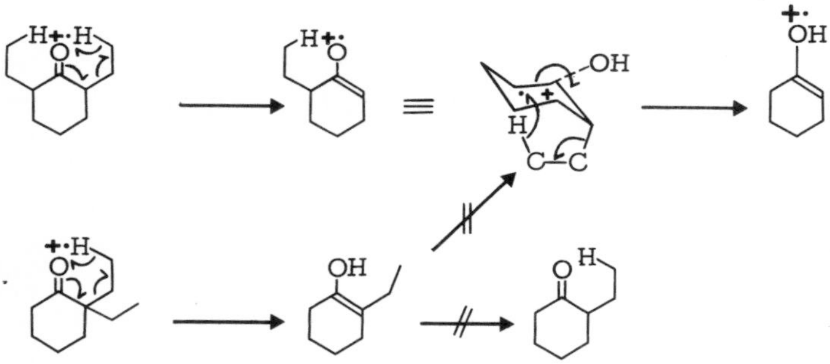

Elimination of a neutral molecule between adjacent groups on a cis-double bond or aromatic system

A neutral molecule can be eliminated by a rearrangement process from a 1,2-disubstituted *cis*-double bond or an *ortho*-disubstituted aromatic system. One of the functional groups must have a transferable hydrogen atom available and the other must contain the other portion of the neutral molecule being eliminated. This is very similar to a McLafferty rearrangement.

$$\left[\begin{array}{c}\text{B-D-R}\\\text{A-H}\end{array}\right]^{+\cdot} \longrightarrow \left[\begin{array}{c}\text{B}\\\text{A}\end{array}\right]^{+\cdot} + \begin{array}{c}\text{D-R}\\|\\\text{H}\end{array}$$

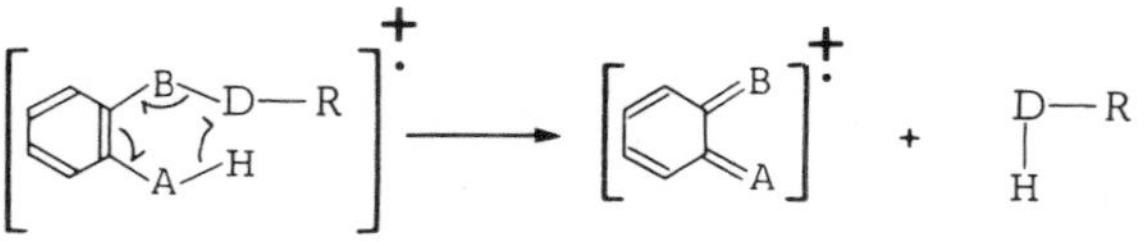

The mechanism involves a six-membered transition state. The atoms A, B and D can be carbon, oxygen, nitrogen, or sulphur. The double bond can be in any system provided that the groups involved are in suitable proximity. The eliminated molecule contains R, which can have a wide range of values. Common examples of the types of molecules

eliminated are water, alcohols, thiols, ammonia, amines and acids,
e.g.

$$\left[\begin{array}{c}\text{[structure with C=O, O-CH}_3\text{, OH]}\end{array}\right]^{+\cdot} \longrightarrow \left[\begin{array}{c}\text{[structure with C=O, =O]}\end{array}\right]^{+\cdot} + \begin{array}{c}\text{O-CH}_3\\|\\\text{H}\end{array}$$

Note again the stabilities of the fragment ion and neutral molecule formed.

With the substituted double bond this provides a means of readily distinguishing between *cis* and *trans* isomers. The *trans* isomer must first convert to the *cis* isomer before rearrangement can occur. The peak corresponding to the rearrangement will, therefore, be of much lower abundance in the case of the *trans* isomer.

Elimination of a neutral molecule containing one of the functional groups in a molecular ion and one hydrogen atom

Elimination of a neutral molecule can occur in a random manner from some molecular ions. X can be ester, hydroxyl, thiol, halogen, and also nitrile in special circumstances.

$$\left[\begin{array}{c}C-X\\(C)_n\\C-H\end{array}\right]^{+\cdot} \longrightarrow \left[\begin{array}{c}C\\(C)_n\\C\end{array}\right]^{+\cdot} + \begin{array}{c}X\\|\\H\end{array}$$

It has been shown by deuteration experiments that this elimination can occur when n has the range of values 0, 1, 2, 3, 4 etc. depending on the environments, e.g. it has been shown that the elimination is mainly 1-3 and 1-4 in 1-butanol, i.e.

$$\left[\begin{array}{c}C-OH\\C\\C-H\\C\end{array}\right]^{+\cdot} \longrightarrow \left[\begin{array}{c}C\\C\\C\end{array}\right]^{+\cdot} \quad \text{and} \quad \left[\begin{array}{c}C-OH\\C\\C\\C-H\end{array}\right]^{+\cdot} \longrightarrow \left[\begin{array}{c}C\\C\\C\\C\end{array}\right]^{+\cdot}$$

Similar behaviour has been shown to occur in cyclohexanol, i.e.

$$[\bigcirc]^{+\cdot} \xleftarrow{1-4} \left[\begin{array}{c}\text{OH}\\\text{cyclohexanol with H at 3,4}\end{array}\right]^{+\cdot} \xrightarrow{1-3} [\bigcirc]^{+\cdot}$$

It must be remembered that this type of elimination from the molecule can occur thermally. In this case the elimination is 1-2 and yields the corresponding olefine. The extent to which this occurs depends on the temperature if a heated inlet system is used. It can be greatly reduced if a direct insertion probe is used because ionization occurs on evaporation of the sample. Conditions must be carefully controlled to obtain reproducible spectra of compounds of this type. When X is halogen this elimination will occur, the extent being governed by the carbon–halogen bond energy. Fluorine and chlorine favour this elimination while bromine and iodine do not.

With esters the acid group can be eliminated by the above mechanism. However it is also possible to eliminate the acyl grouping as a stable neutral molecule (a ketene) when no hydrogen is available in the alcohol grouping, e.g. benzyl acetate:

$$\left[\text{C}_6\text{H}_5\text{—CH}_2\text{—O} \overset{\text{O}}{\underset{\text{H}}{-\text{C}-}} \text{CH}_2 \right]^{+\cdot} \longrightarrow \left[\text{C}_6\text{H}_5\text{—CH}_2\text{OH} \right]^{+\cdot} + \text{CH}_2\text{=C=O}$$

With esters the McLafferty rearrangement described above is possible, and when the conditions mentioned are fulfilled this process will compete. The same is true for the nitrile group which will only undergo the elimination described above if it cannot for some reason decompose by a McLafferty rearrangement. Sulphur analogues of esters and alcohols behave in a similar manner, but to a lesser extent, and nitrogen containing groups usually direct so strongly in other directions that this type of fission process is not favoured.

Double rearrangement processes
Sometimes peaks are observed in a mass spectrum which can only be explained by simultaneous rearrangement of two hydrogen atoms from the group being eliminated to the fragment ion being formed. This ion will, therefore, be two mass units heavier than a corresponding fragment ion formed by simple fission.

Esters greater than methyl and carbonates greater than methyl
The molecules under consideration have the following general formula:

$$\text{R—C}\underset{\text{O}\dagger\text{CH}_2\text{—(A)}_n\text{—BH}}{\overset{\text{O}}{\diagup\!\!\!\diagdown}}^{\text{F}}$$

Simple fission at F would yield an ion $[\text{RCO}_2]^+$. Normal McLafferty rearrangement could lead to an ion

$$\text{R—C}\underset{\text{OH}}{\overset{\text{O}^{\cdot+}}{\diagup\!\!\!\diagdown}}$$

and double rearrangement to an ion

$$\text{R—C}\underset{\text{OH}}{\overset{\text{OH}^+}{\diagup\!\!\!\diagdown}}$$

As can be seen, the ion formed from the double rearrangement is a protonated acid species and is very stable. The odd-electron neutral fragment also has considerable stability in every case due to resonance stabilization. Since this process applies to almost every ester greater than methyl its existence should be noted. A possible mechanism is as follows:

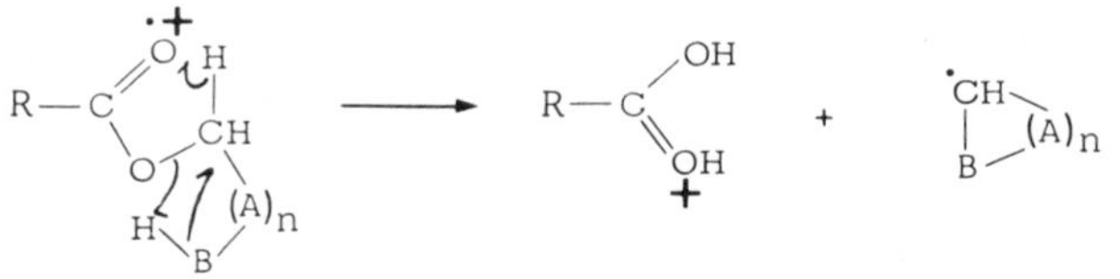

R can be almost any system including OR' when the molecule is a carbonate. A can be carbon, sulphur or oxygen. n, which can have values 0, 1, 2, 3 etc., applies to combinations which are chemically feasible e.g. $CH_2CH_2CH_2$, CH_2OCH_2 or CH_2CH_2O are all reasonable for $(A)_n$ where $n = 3$. B can also be carbon, oxygen or sulphur with the restriction that the combination $-(A)_n BH$ must be realistic. It has not been proved exactly from which two centres the hydrogen atoms come but there must be at least two with hydrogen available. The fragment ion produced immediately indicates the size of the acid side of the ester, and the fact that the rearrangement has occurred gives information about the alcohol side. The peak at m/e 61 in the mass spectra of acetates is a common example, and for cases when $R \equiv C_n H_{2n+1}$ the peak will occur at m/e $(61 + n \times 14)$ e.g. the peak at m/e 89 in butanoate etc. In come cases, because of competition with other processes, the abundance of the peak due to double rearrangement is very low. It can, however, be one of the major peaks in the mass spectrum under favourable conditions. Some examples of this rearrangement, selected from a vast number, are as follows:

di-n-propyl carbonate

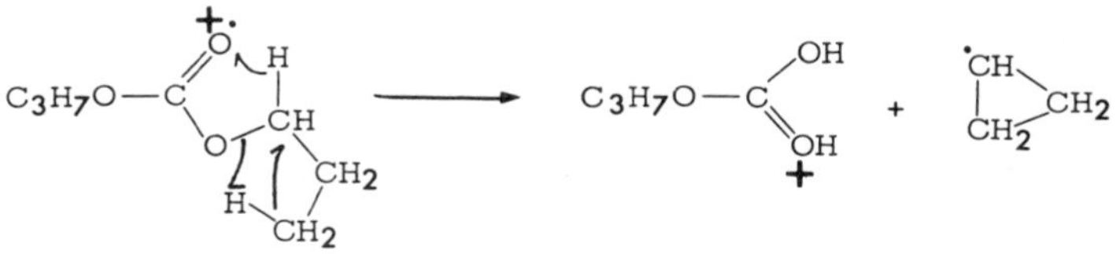

m/e 146 m/e 105

In the mass spectrum of this compound the ion m/e 105 is the highest mass peak which is significant in size.

β-hydroxy-p-anisate

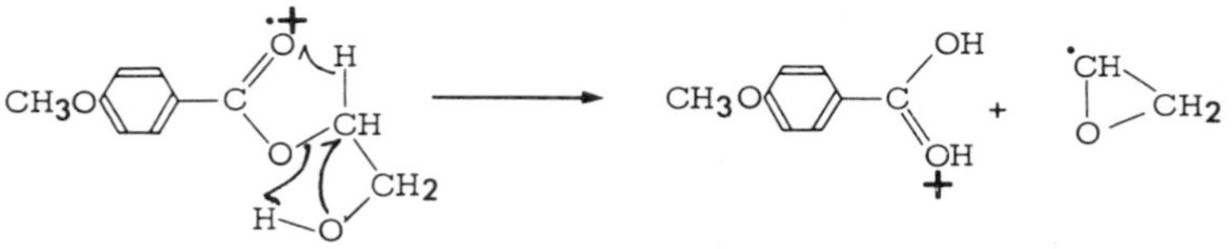

m/e 196 m/e 153

n-butyl furoate

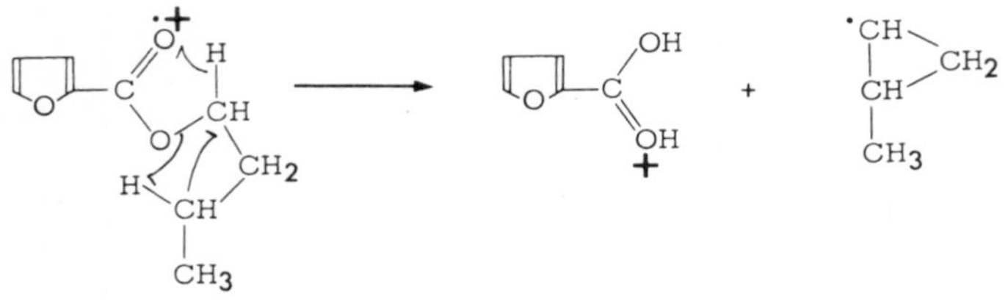

m/e 168 → m/e 113

2-chloroallyl acetate

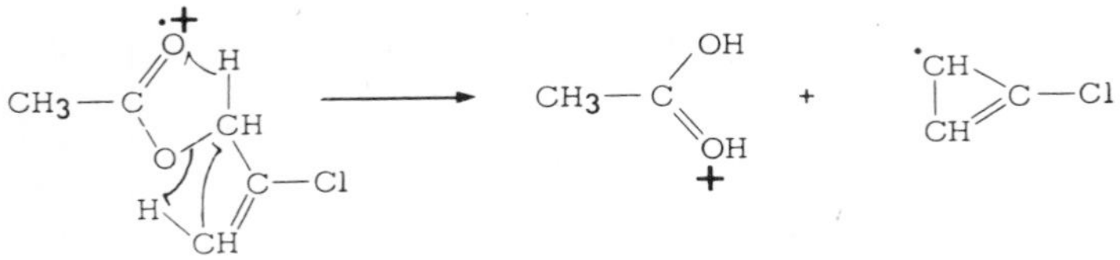

m/e 134 → m/e 61

isopropyl butanoate

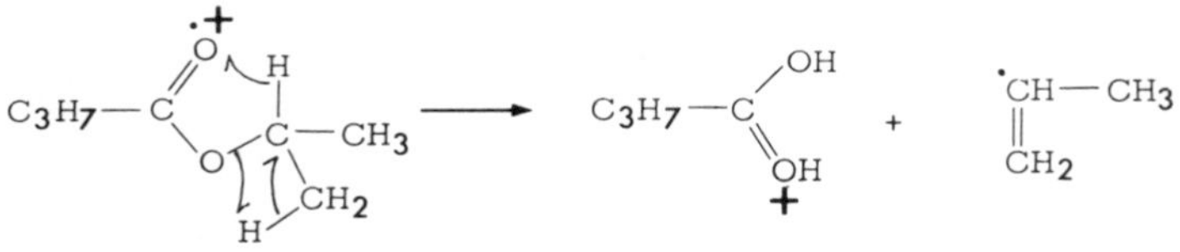

m/e 130 → m/e 89

diethyl malonate

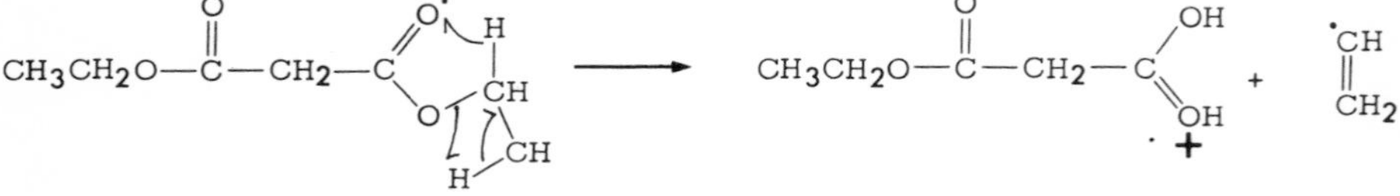

m/e 160 → m/e 133

Suitable functional groups substituted on adjacent carbon atoms
In the case of ethylene glycol an abundant peak is observed in the mass spectrum at *m/e* 33 and can be formed only by rearrangement of two hydrogen atoms. McLafferty[8] has

proposed the following mechanism which involves two four-membered transition states acting concertedly:

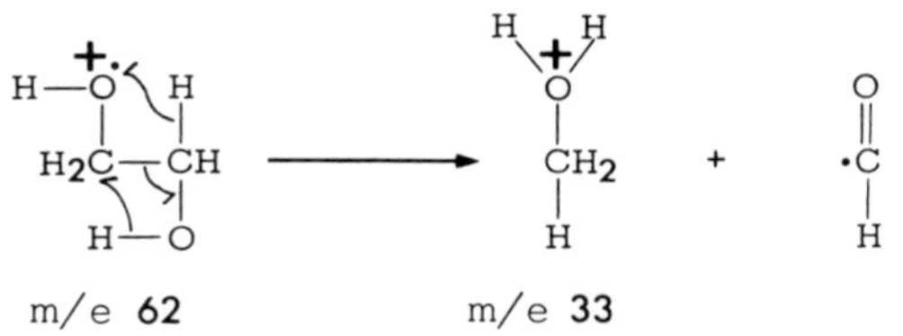

$$\underset{m/e\ 62}{\begin{array}{c}H-\overset{+\cdot}{O}\diagdown H\\|\quad\ |\\H_2C\!-\!CH\\|\quad\ |\\H\!-\!O\end{array}} \longrightarrow \underset{m/e\ 33}{\begin{array}{c}H\diagdown\overset{+}{\diagup}H\\O\\|\\CH_2\\|\\H\end{array}} + \begin{array}{c}O\\\|\\\cdot C\\|\\H\end{array}$$

The resulting ion m/e 33 is a stable protonated species. This mechanism will probably operate for the sulphur analogue and for the mixed thiol-alcohol. By formulating a general mechanism as shown and checking through a number of spectra it can be seen that this

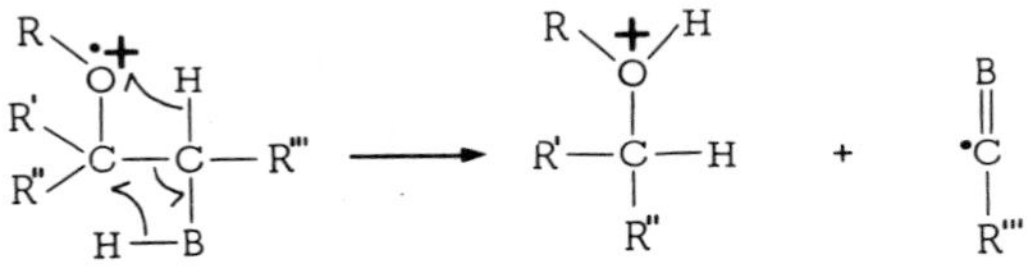

$$\begin{array}{c}R\diagdown\overset{+\cdot}{O}\diagdown H\\R'\quad|\quad\ |\\R''\!\!\!>\!\!C\!-\!C\!-\!R'''\\|\quad\ |\\H\!-\!B\end{array} \longrightarrow \begin{array}{c}R\diagdown\overset{+}{\diagup}H\\O\\|\\R'\!-\!C\!-\!H\\|\\R''\end{array} + \begin{array}{c}B\\\|\\\cdot C\\|\\R'''\end{array}$$

mechanism appears to operate in a large number of cases. B can be CH_2, O or S. When two different functional groups which fulfil the above conditions are present, the factors governing which of the two gives rise to the most abundant peak are the stabilities of the fragment ions and neutral fragments formed. Some examples of peaks which appear to arise by this mechanism are as follows:

2-methoxyethanol

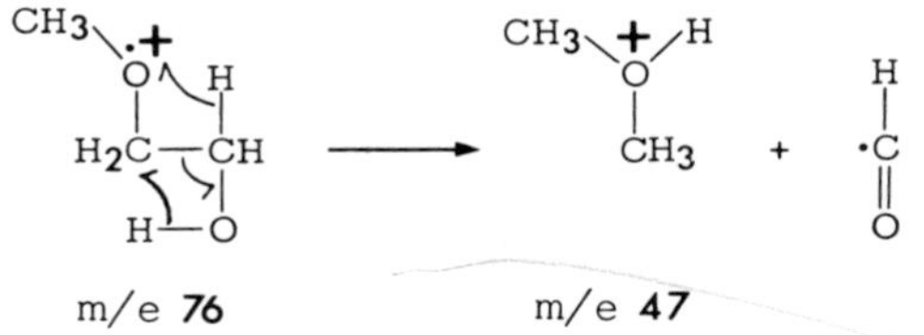

$$\underset{m/e\ 76}{\begin{array}{c}CH_3\diagdown\overset{+\cdot}{O}\diagdown H\\|\quad\ |\\H_2C\!-\!CH\\|\quad\ |\\H\!-\!O\end{array}} \longrightarrow \underset{m/e\ 47}{\begin{array}{c}CH_3\diagdown\overset{+}{\diagup}H\\O\\|\\CH_3\end{array}} + \begin{array}{c}H\\|\\\cdot C\\\|\\O\end{array}$$

Epoxides can also produce an ion by double rearrangement e.g.

1,2-epoxybutane

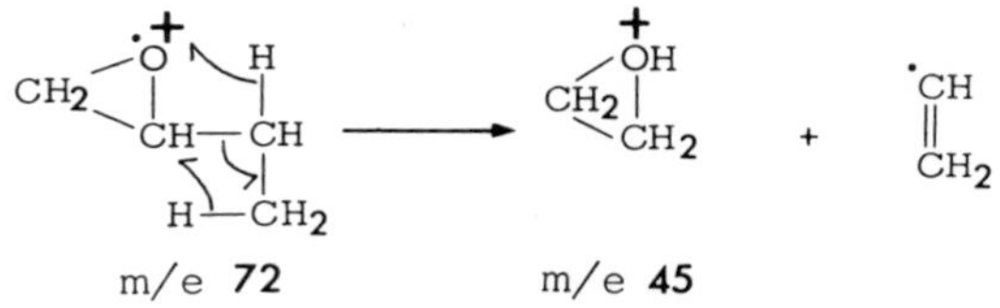

$$\underset{m/e\ 72}{\begin{array}{c}CH_2\diagdown\overset{+\cdot}{O}\diagdown H\\|\qquad\ |\\CH\!-\!CH\\|\quad\ |\\H\!-\!CH_2\end{array}} \longrightarrow \underset{m/e\ 45}{\begin{array}{c}\overset{+}{OH}\\CH_2\diagup|\\\diagdown CH_2\end{array}} + \begin{array}{c}\cdot CH\\\|\\CH_2\end{array}$$

isobutanol

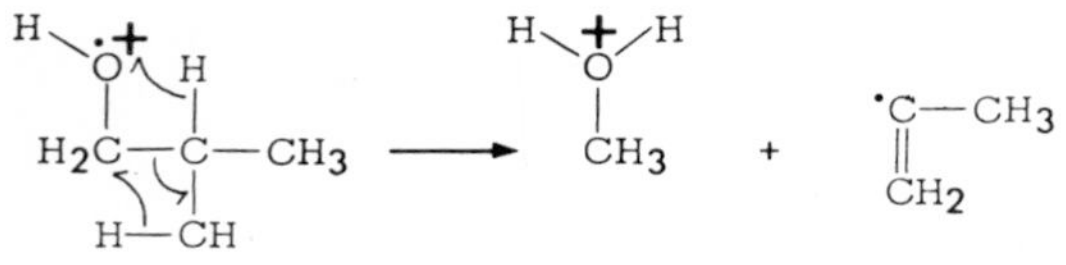

$$\text{[structure: } H_2C\text{—}C(\text{CH}_3)(\text{H-CH})\text{—with } \overset{+}{O}H \cdot H] \longrightarrow [\overset{H\ +\ H}{O}\text{-CH}_3] + [\overset{\cdot}{C}\text{—CH}_3 \parallel \text{CH}_2]$$

m/e 74   m/e 33

It is, therefore, worthwhile remembering that this type of double rearrangement exists if peaks appear in the mass spectrum which cannot be rationalized using simple fission or normal rearrangement processes.

Although the processes are not completely rationalized, some very stable ions appear in mass spectra due to double rearrangement. The two types discussed above are not the only possible examples of this behaviour.

Elimination of a molecule of hydrogen
During the fragmentation sequences occurring in some complex molecules some fragment ions can be seen to have eliminated a molecule of hydrogen. Aromatization of the fragment ion is usually the driving force for this process. It is difficult to state a rule to explain this, but a four-membered transition state can be envisaged. Metastable ions often prove that loss of hydrogen has occurred e.g. many ions of m/e 93 ($= C_7H_9$) lose hydrogen presumably to form the stable aromatic tropylium ion. The following ion derived from ibogaine[5] eliminates a hydrogen molecule:

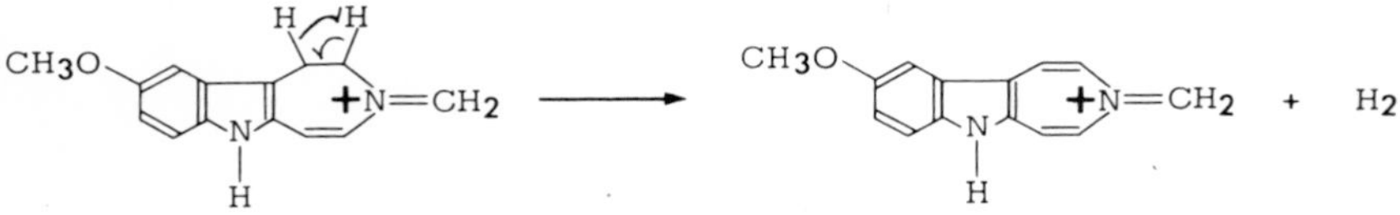

$$\text{CH}_3\text{O—[indole with } \overset{+}{N}=\text{CH}_2 \text{ and H—H group]} \longrightarrow \text{CH}_3\text{O—[aromatized indole with } \overset{+}{N}=\text{CH}_2] + H_2$$

Rearrangements occurring in fragment ions
Some rearrangement processes only occur from suitably substituted fragment ions formed by simple fission processes:

1. With hydroxyl, thiol, and amino groups on primary, secondary or tertiary carbon atoms the favoured simple fission process was discussed earlier i.e. α-fission.

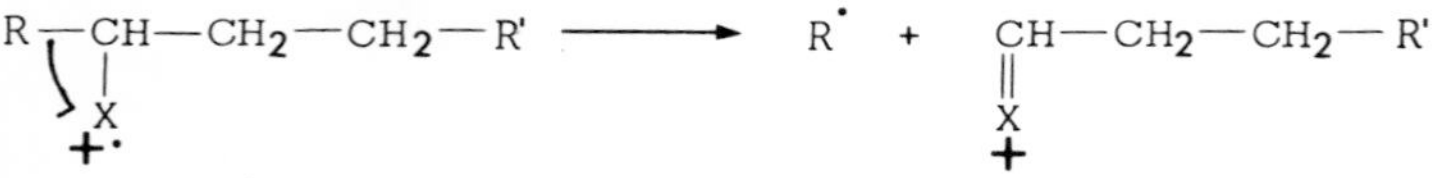

$$R\text{—CH}(X)\text{—CH}_2\text{—CH}_2\text{—R'} \longrightarrow R^{\cdot} + \overset{+}{\text{CH}}(=X)\text{—CH}_2\text{—CH}_2\text{—R'}$$

The ion formed, which is stabilized by resonance can, if there is a side chain containing a radical larger than methyl, eliminate an olefine molecule as shown below:

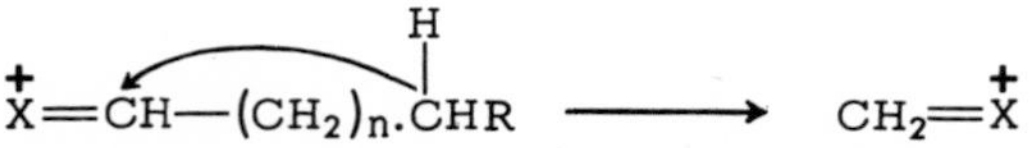

$$\overset{+}{X}=\text{CH—(CH}_2)_n.\text{CHR(H)} \longrightarrow \text{CH}_2=\overset{+}{X} \qquad (n \neq 0)$$

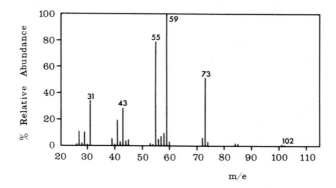

FIG. 2. Mass spectrum of hexan-3-ol.

The spectrum of hexan-3-ol shown in Fig. 2 has a peak at m/e 31 resulting from this process. This spectrum also illustrates that the most prominent of the α-cleavages which occur in the molecular ion is the one corresponding to the loss of the larger radical. The sequence m/e 102 → m/e 73 → m/e 31 also occurs by the mechanism shown below, but starting with the elimination of the ethyl group.

The ion m/e 31 is the ion which is formed by simple fission in a primary alcohol molecular ion i.e.

This does not cause confusion because, being a derived ion when it is formed by rearrangement, it is of much lower abundance than the ion from which it is formed.
2. When simple fission occurs in the molecular ion of an ether, thioether, or a secondary or tertiary amine, the 'stable' onium ion thus produced can eliminate an olefine molecule by rearrangement if the alkyl group left is more than two carbon atoms long. α-fission

The rearrangement is as follows:

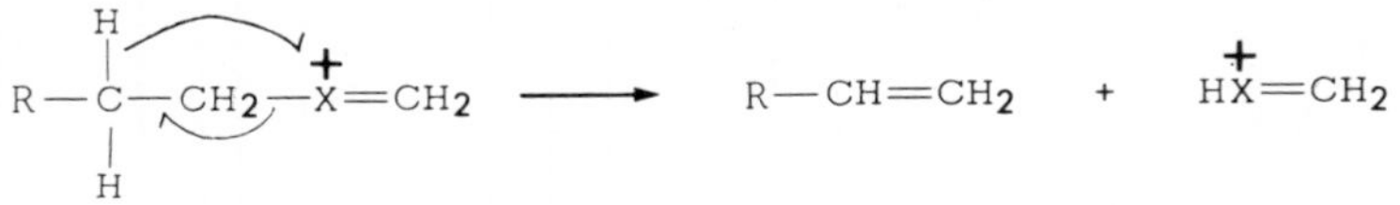

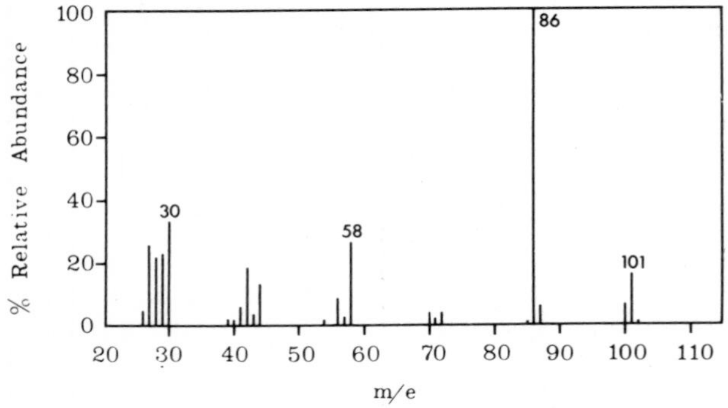

$$R-\underset{\underset{H}{|}}{\overset{\overset{H}{|}}{C}}-CH_2-\overset{+}{X}=CH_2 \longrightarrow R-CH=CH_2 + H\overset{+}{X}=CH_2$$

FIG. 3. Mass spectrum of triethylamine.

To illustrate this, the case of triethylamine $(CH_3CH_2)_3N$ may be considered (see Fig. 3). The mechanism probably operates successively as follows:

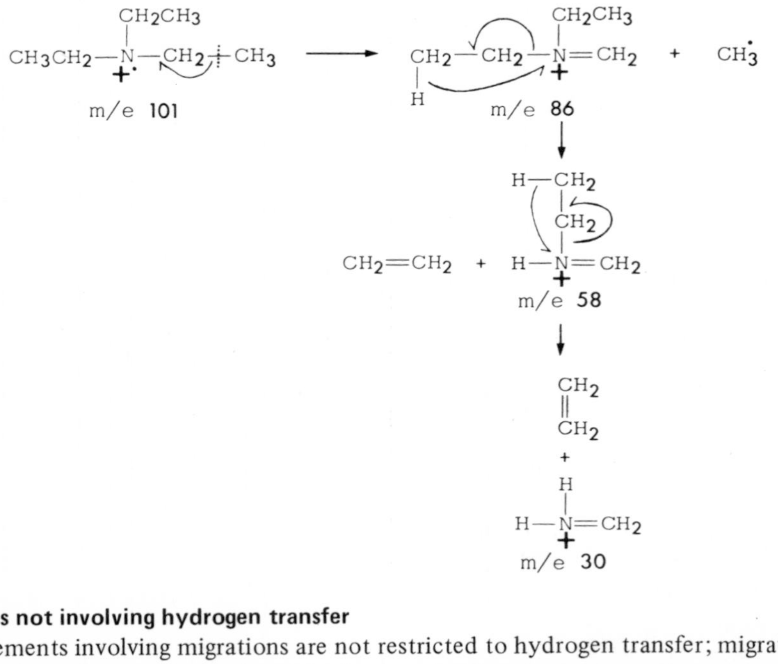

$$CH_3CH_2-\underset{\underset{+\cdot}{|}}{\overset{\overset{CH_2CH_3}{|}}{N}}-CH_2-CH_3 \longrightarrow \underset{\underset{H}{|}}{\overset{\overset{CH_2CH_3}{|}}{CH_2-CH_2-\overset{+}{N}=CH_2}} + CH_3^{\cdot}$$

m/e 101   m/e 86

$$CH_2=CH_2 + H-\overset{+}{N}=CH_2$$
m/e 58

$$\underset{\underset{\underset{\underset{m/e\ 30}{}}{H-\overset{+}{N}=CH_2}}{\overset{H}{|}}}{\overset{\overset{CH_2}{\|}}{CH_2}}$$

Migrations not involving hydrogen transfer

Rearrangements involving migrations are not restricted to hydrogen transfer; migrations of aryl, methyl, methoxy, ethyl, ethoxy, hydroxy and other groups are quite common. The presence of these migrations can complicate the process of structural deduction, particularly as there are no general rules available to predict or recognize the occurrence of such rearrangements in the mass spectrum of an unknown molecule. A few examples, chosen to show the wide variety of such rearrangements, are given below. Cooks in his review[9] gives many more.

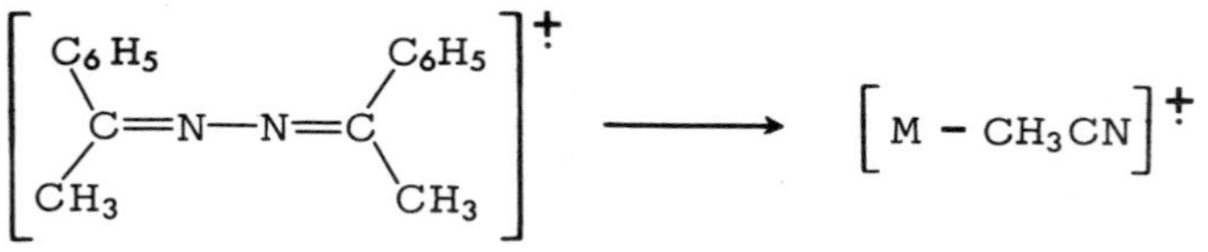

$$\left[\begin{array}{c} C_6H_5 \\ C=N-N=C \\ CH_3 \end{array} \begin{array}{c} C_6H_5 \\ CH_3 \end{array} \right]^{+\cdot} \longrightarrow \left[M - CH_3CN \right]^{+\cdot}$$

In the example above a phenyl migration presumably occurs, whereas the case below is probably one of a methyl or ethyl migration since there is no $[M - CO_2]^+$ ion in the spectra of those compounds in which R = propyl.

$$\left[R'R''CH_2CH(CN)COOR \right]^{+\cdot} \longrightarrow \left[M - CO_2 \right]^{+\cdot}$$

where R = methyl or ethyl only.

In the case of the loss of carbon monoxide from 2-benzoylthiophene it is not clear whether this is a phenyl or thiophene radical migration.

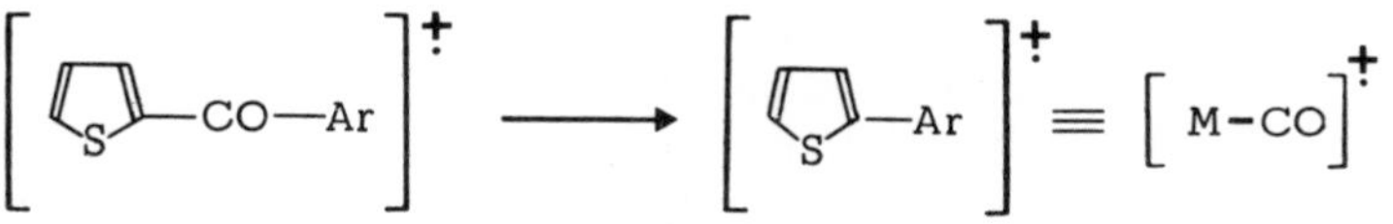

$$\left[\underset{S}{\bigcirc}-CO-Ar \right]^{+\cdot} \longrightarrow \left[\underset{S}{\bigcirc}-Ar \right]^{+\cdot} \equiv \left[M-CO \right]^{+\cdot}$$

Similarly with arylsulphonylchlorides, it is not clear whether the $[M - SO_2]^+$ ion comes from the migration of the chlorine or the aryl group, but on the basis of the lower bond energy of the sulphur–chlorine bond the former may be the case:

$$\left[RC_6H_4SO_2Cl \right]^{+\cdot} \longrightarrow \left[RC_6H_4Cl \right]^{+\cdot}$$

An example of methoxy migration is shown below, and migrations of ethoxy groups have also been reported:

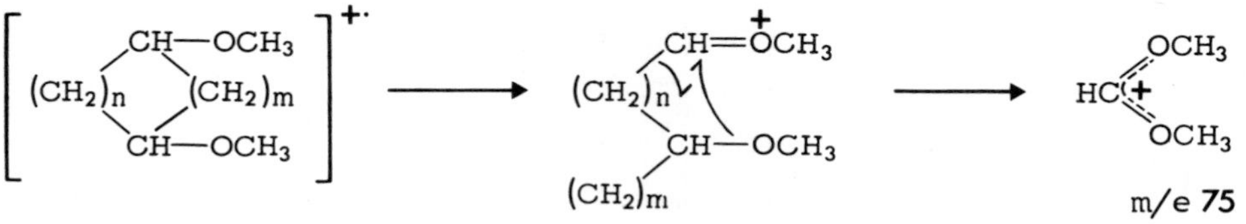

$$\left[\begin{array}{c} CH-OCH_3 \\ (CH_2)_n \quad (CH_2)_m \\ CH-OCH_3 \end{array} \right]^{+\cdot} \longrightarrow \begin{array}{c} CH=\overset{+}{O}CH_3 \\ (CH_2)_n \\ CH-OCH_3 \\ (CH_2)_m \end{array} \longrightarrow HC\overset{+}{\underset{OCH_3}{\overset{OCH_3}{\diagup}}}$$

m/e 75

where $n = 1$ to 5 and $n + m = 3, 4$ or 5.

Although the migration of these groups is less frequent than hydrogen, their possible use for structural determination by mass spectrometry raises serious problems compared with hydrogen migrations which are often of use in determining structures.

COMPLEX FISSION PROCESSES

In this section some of the processes by which rings fragment are discussed. When a simple fragmentation of a ring bond happens, due to one of the factors discussed on p. 35, no elimination of a neutral fragment can occur without further cleavage, as borne out by the fragmentation patterns of various cyclic systems shown below.

Saturated cyclic systems

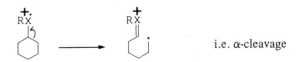

i.e. α-cleavage

(X = NR′, S or O).

For elimination of a neutral fragment, a second bond fission, *with* or *without* hydrogen transfer, must occur. The number of bonds broken and hydrogen atoms transferred depends upon the structure involved, and the usual stability considerations about the resulting fragment ion and neutral molecule or fragment. The ion which is formed by this process usually has high stability. There are too many different environments possible to allow discussion of every case. A few examples will, however, illustrate the general approach.

When attempting to rationalize the genesis of a fragment ion of this kind the first step is to find the most probable sites which will compete for the localization of the positive charge. The next step is to apply the necessary fragmentation mechanisms to the structure to find which of the potential fragment ion—neutral molecule combinations is energetically most favourable. The relative abundances of peaks in the mass spectrum corresponding to the ions formed by applying the same mechanism along different paths is an indication of the favourability of each path. With experience the path which will yield the most stable ion can be correctly selected from several possibilities. When two or more functional groups are present, or more than one mechanism is possible for a single functional group, fragment stability again decides the order of precedence of these. This interplay can accentuate, or completely suppress, the formation of an ion which at first sight should be formed.

The difference between a complex fragmentation process and the multi-step fragmentation of a molecular ion is that in the latter case a metastable ion can (if favourable) be observed for each step. With a complex fragmentation only the initial and final ions involved are recorded. The intermediate steps have been postulated and in many cases confirmed by isotope labelling experiments.

In the case of the cyclic molecular ion shown above after one bond fission the next steps are as follows:

      A   B   C

The hydrogen atom indicated is transferred as shown in A. The primary free radical in A is thus changed to the more stable secondary free radical shown in B. The most vulnerable bond (1) then breaks as shown in B to yield a stable ion C, and a neutral fragment. With the six-membered cyclic system discussed above it is also possible to envisage the above fragment ion C being formed from the molecular ion A by means of a six-membered transition state as a concerted process, i.e.

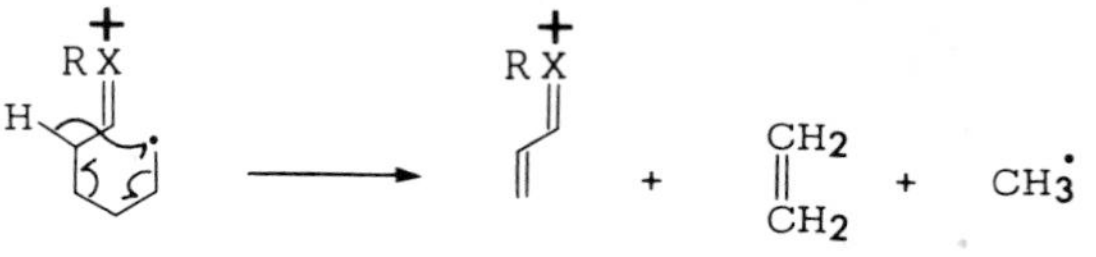

Because only the ion C is recorded as a peak in the mass spectrum either mechanism could operate. However, labelling shows that it is the hydrogen on C-2 which is transferred.

In the section on pretreatment, ethylene ketals, ethylene thioketals and N,N-dimethylamine derivatives were mentioned because the ion which is formed when any of these functional groups is present in a molecule is so stable that it completely dominates the fragmentation of the molecule. The mechanism, the first stage of which is essentially similar to the one outlined for the cyclohexane derivatives (A → B → C), has been confirmed by deuterium labelling of the steroidal ethylene ketal derivative below. As in the case of 2-methylcyclohexanone, cleavage of the two α-bonds leads to two different ions.

Fission of bond (1)

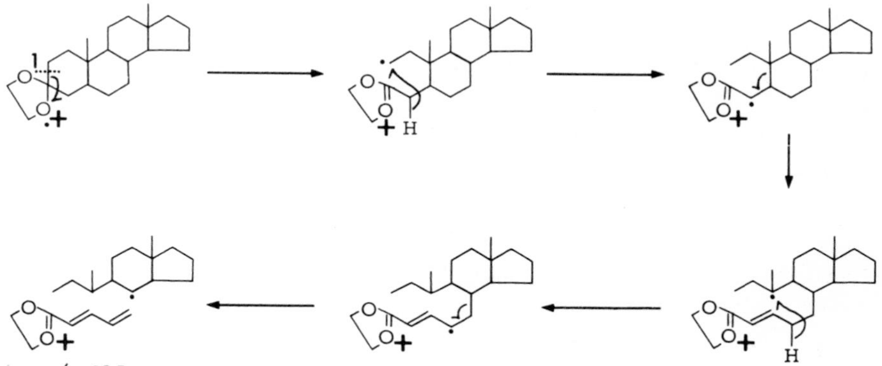

m/e 125

Fission of bond (2)

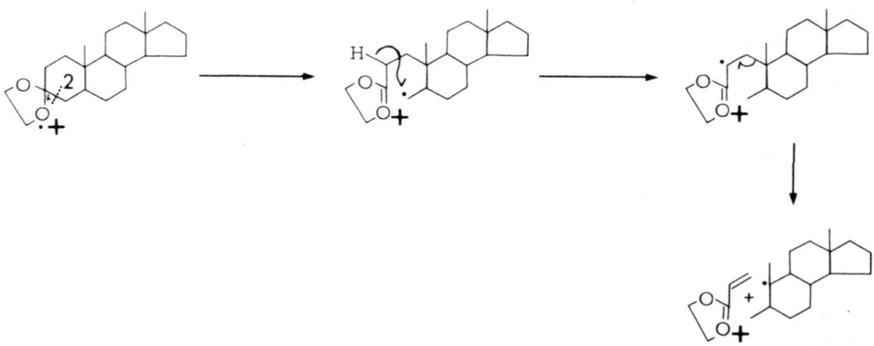

m/e 99

Any substitution in the part of the nucleus which becomes part of the fragment ion will increase the fragment ion mass accordingly. Theoretically, any cyclic carbonyl-containing compound which fulfils the requirements will, on conversion to the ketal derivative, yield two fragment ions as shown above by fission of bonds (1) and (2). If substitution in the cyclic system makes one of the fragment ions more stable than the other, the abundance of the corresponding peak in the mass spectrum will reflect this. The interplay of stability and ring bond lability was discussed for this type of process in the lability section. The corresponding ions for the N,N'-dimethyl derivatives are:

Bond (1)   Bond (2)

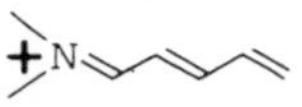

m/e 110   m/e 84

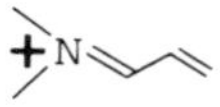

and for the ethylene thioketal, due to the replacement of two oxygen atoms by sulphur, the m/e values of the basic ions increase by 32 i.e. m/e 157 and 131.

Although this mechanism is essentially the same as that which gives rise to m/e 55 in cyclohexanone (see p. 39), in larger molecules the carbonyl group does not seem so effective in producing an intense ion corresponding to this fragmentation. This is why the ethylene ketal derivatives of the carbonyl are used.

This type of fragmentation can be suppressed by structural modifications in all cases. Thus, in the unsaturated ketone shown below, the double bond prevents this fragmentation and the alternative fragmentation shown takes place. In the case of the ethylene ketals

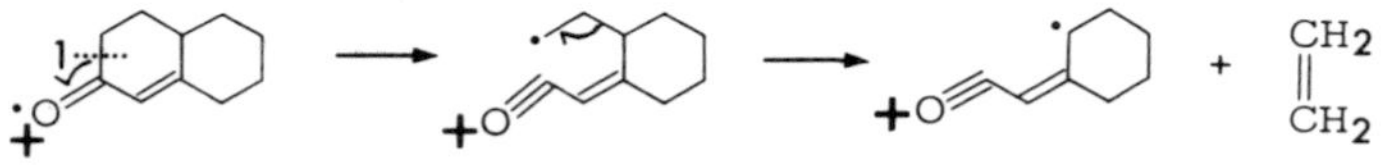

described above the presence of two methyl groups at C-4, as are often found in triterpenes, suppresses the fragmentation pathway leading to m/e 125 since there is no hydrogen available for the first transfer step. There is no ion at m/e 139 which shows that methyl groups do not transfer under these conditions, nor does cleavage of the C-3–C-4 bond predominate due to the stabilizing powers of the two methyl groups on C-4 (see p. 38).

The retro-Diels–Alder reaction
This reaction occurs in very many types of molecules thermally, photochemically and under electron impact. The simplest example is the formation of the $[M - C_2H_4]^{+\cdot}$ ion in the mass spectrum of cyclohexene:

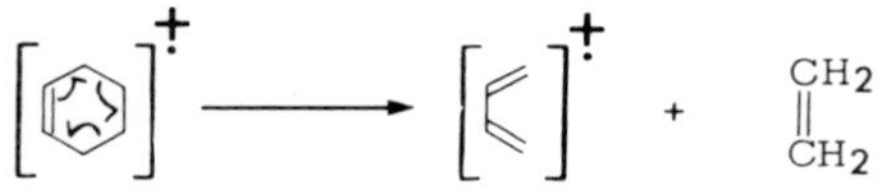

The recognition of the original structure of a molecule from ions corresponding to this fragmentation can sometimes be complicated by the fact that under electron impact the hydrogens and hence the 'double bond' in a cyclic olefine migrate in the molecular ion before fragmentation. Thus in a compound of formula C_8H_{14} the most intense even m/e ion is at m/e 68. This would suggest that the compound is 3,4-dimethylcyclohexene undergoing a retro-Diels–Alder fragmentation. However, the compound is found to be 1,2-di-

$$\left[\begin{array}{c} \end{array} \right]^{\ddagger} \longrightarrow \left[\begin{array}{c} \end{array} \right]^{\ddagger}$$
$$m/e\ 68$$

methylcyclohexene and to give m/e 68 by a retro-Diels–Alder reaction the double bond must first have isomerized. Indeed, in the spectrum of this compound ions are seen corresponding to retro-Diels–Alder reactions of all the possible isomers. The fact that

$$\left[\begin{array}{c} \end{array} \right]^{\ddagger}_{1,2} \rightleftharpoons \left[\begin{array}{c} \end{array} \right]^{\ddagger}_{2,3} \rightleftharpoons \left[\begin{array}{c} \end{array} \right]^{\ddagger}_{3,4} \rightleftharpoons \left[\begin{array}{c} \end{array} \right]^{\ddagger}_{4,5}$$

$$\downarrow \qquad \downarrow \qquad \downarrow \qquad \downarrow$$

$$\left[\begin{array}{c} \end{array} \right]^{\ddagger} \quad \left[\begin{array}{c} \end{array} \right]^{\ddagger} \quad \left[\begin{array}{c} \end{array} \right]^{\ddagger} \quad \left[\begin{array}{c} \end{array} \right]^{\ddagger}$$

m/e 68 is the most intense of these ions has been related to the idea that the retro-Diels–Alder electron impact reaction goes via a linear transition state as shown below. In the above case the most stable ion would be the one which would give m/e 68, since both the charge and radical are located on tertiary sites. Whether the retro-Diels–Alder reaction under electron impact is concerted or stepwise is still a matter of doubt.

$$\left[\begin{array}{c} \end{array} \right]^{\ddagger} \longrightarrow \begin{array}{c} \end{array} \longrightarrow m/e\ 68$$

In other cases the migration of the double bond leads to an ion which cannot fragment any further. In this case the type of confusion which has been discussed above does not exist and the retro-Diels–Alder ion corresponding to the original structure is prominent and can be used to determine the structure of the molecule. This is so in the case of Δ^{12}-oleanene and similar pentacyclic triterpenes having the Δ^{12} double bond.[7] Further fragmentation of the retro-Diels–Alder ion can sometimes help in structure determination. The sequence shown below led to the partial structure determination and identification with a known compound. The π-electrons necessary for this process can be those of an

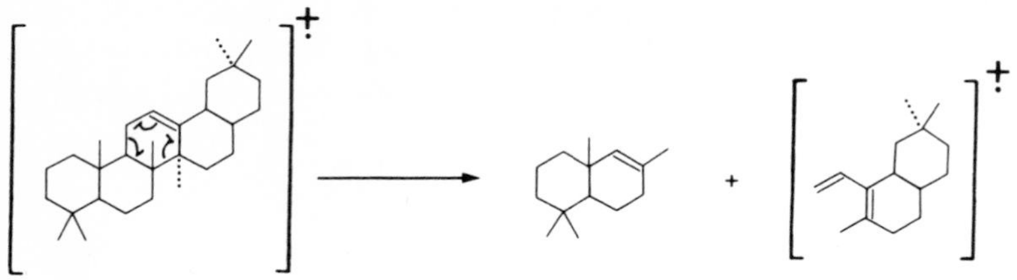

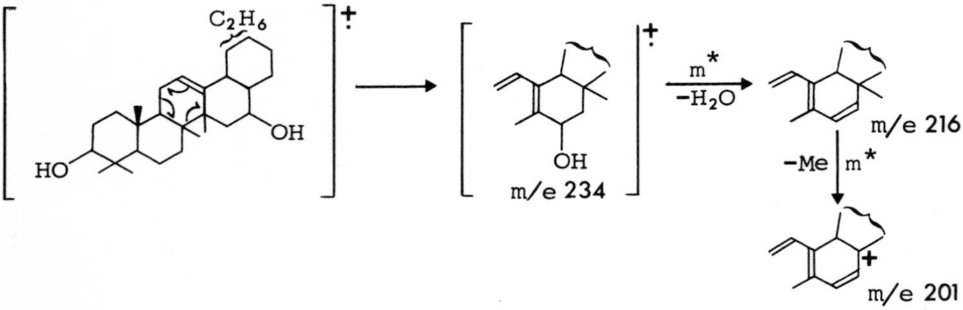

aromatic nucleus as, for instance, in the formation of the ion at m/e 104 in the spectrum of isochroman. This reaction is often important in the determination of the structure of

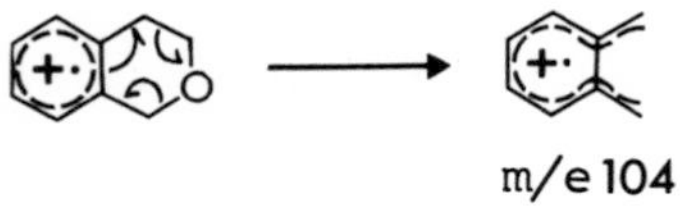

some alkaloids.[4] Thus, in the case of the crotonosine alkaloids the presence or absence of a N-methyl grouping could be determined by the retro-Diels–Alder reaction shown below. In the case of R = H, high resolution mass measurement is necessary to distinguish between the $[M - CHO]^+$ and retro-Diels–Alder ion $[M - CH_2=NH]^+$, both of which occur at m/e M − 29.

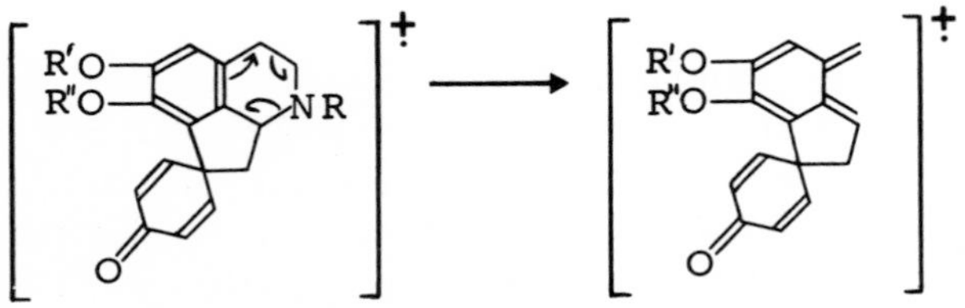

Complex fragmentation in aromatic systems

In aromatic molecules which have no easily fragmentable side chain many complex reactions involving fragmentation of the ring occur. The mechanism of these reactions is not well understood, although in the first two cases it is known that no cleavage of the heteroatom to carbon bond, originally present, happens. This means that ring expansion, as in the toluene molecular ion, does not occur. Some examples are:

$$[C_6H_5OH]^{+\cdot} \longrightarrow [M-CO]^{+\cdot}$$

$$[C_6H_5NH_2]^{+\cdot} \longrightarrow [M-HCN]^{+\cdot}$$

$$[C_6H_5NSO]^{+\cdot} \longrightarrow [M-CO]^{+\cdot}$$

FORMATION OF A MASS SPECTRUM

So far in this chapter various types of fragmentation processes have been discussed. A mass spectrum is made up of several of these processes, often either in competition or following one another. When an ion breaks down further it must do so in an energetically favourable manner. At the present state of knowledge it is not possible to set down rules to indicate which processes will occur consecutively when an ion is fragmenting. However, with simple molecules common breakdown sequences have one simple fission process (S) in them. It can be preceded or followed by one or more rearrangement process (R), i.e.

(a) $\quad M^{+\cdot} \xrightarrow{(S)} F^{+} \xrightarrow{(R)} F_1^{+} \xrightarrow{(R)} F_2^{+}$

(b) $\quad M^{+\cdot} \xrightarrow{(R)} F_3^{+\cdot} \xrightarrow{(S)} F_4^{+} \xrightarrow{(R)} F_5^{+}$

(c) $\quad M^{+\cdot} \xrightarrow{(R)} F_6^{+\cdot} \xrightarrow{(R)} F_7^{+\cdot} \xrightarrow{(S)} F_8^{+}$

The subscripts 1–8 merely indicate different fragment ions. In particular instances some of these ions will have the same m/e value or formula. More than one degradation sequence of the same type [e.g. (a)] may occur for one molecular ion and will lead to ions of different mass being formed when the neutral fragments eliminated are of different mass. It is by competing degradation sequences that the large number of ions corresponding to peaks in the mass spectrum are formed. Metastable peaks, when observed, are used to relate members of the same sequence.

An example which illustrates multi-step fragmentation in the sequence (S)–(R)–(R) is shown by metastable peaks in the mass spectrum of 3-ethyl-3-methylpentane. Both the rearrangements involve at least a hydrogen migration:

$$M^{+\cdot} \xrightarrow[(S)]{-C_2H_5} [M-29]^{+} \xrightarrow[(R)]{-C_2H_4} [M-57]^{+} \xrightarrow[(R)]{-CH_4} m/e\ 41$$

m/e 114 $\qquad\qquad\quad$ m/e 85 $\qquad\qquad\quad$ m/e 57

An example of the sequence (R)–(R)–(S) has been given on p. 66 in the case of the triterpene of unknown structure.

An example of the sequence (R)–(S)–(R) is found in the mass spectrum of n-octane which has been confirmed by metastable evidence.

$$M^{+\cdot} \xrightarrow{(R)} [M-30]^{+} \xrightarrow{(S)} [M-59]^{+} \xrightarrow{(R)} m/e\ 29$$
$$m/e\ 114 \qquad\quad m/e\ 70 \qquad\qquad m/e\ 55$$

More than one simple fission process can occur in a sequence. A common example of this is the elimination of a neutral molecule of carbon monoxide from ions of the type RCO^+ which are formed by a simple fission process, i.e.

$$R-\overset{+\cdot}{\underset{\parallel}{C}}-R' \longrightarrow R-\overset{+}{\underset{\parallel\parallel}{C}} \longrightarrow R^{+} + CO + R'$$

where R can be aromatic or alkyl and R' can be H, alkyl, O-alkyl, aromatic, O-aromatic etc.

It is possible also for other neutral molecules and stable radicals to be eliminated in fragmentation sequences of this type, e.g.

$$\left[\underset{Br}{\overset{Br}{\bigcirc}}\right]^{+\cdot} \xrightarrow{(S)} \left[\overset{Br}{\bigcirc}\right]^{+} \xrightarrow{(S)} [C_6H_4]^{+\cdot}$$

and

$$\left[CH_3-\overset{O}{\underset{\parallel}{C}}-\underset{Cl}{\overset{NO_2}{\bigcirc}}\right]^{+\cdot} \xrightarrow[-CH_3^{\cdot}]{(S)} \left[\overset{O\ H}{\underset{H\ H}{\cdot C}}\underset{Cl}{\overset{NO_2}{\bigcirc}}\right]^{+} \xrightarrow[-NO_2^{\cdot}]{(S)} [CH_3Cl\,CO]^{+\cdot}$$

$$(S)\Big\downarrow -CO$$

$$[C_6H_3]^{+\cdot} \xleftarrow[-Cl^{\cdot}]{(S)} [C_6H_3Cl]^{+\cdot}$$

The main reason for this type of behaviour is that the molecule presents no alternative, more energetically favourable, mode of fission. In addition, the ions formed and the neutral fragments are stable entities.

When cyclic or aromatic systems are part of a molecule, complex fission processes become necessary to eliminate neutral fragments, the component atoms of which form part of the ring system. These processes are less energetically favourable, usually occurring after simple fission processes, and rearrangements involving the non-aromatic or non-cyclic parts of the molecule, have occurred. When complex fission processes are occurring it is useful to look for potential fragment ions which have high stability as a result of aromaticity etc. and for neutral fragments of high stability.

Thus, a mass spectrum is made up of several such reaction sequences. Many examples could be given, but the one below illustrates how several ions are formed from a molecular ion by three competing fragmentation sequences. The molecule is diethyl acetal.

In the next Chapter examples of interpretation of spectra involving several of these sequences are given.

4

Interpretation of the mass spectrum

INTRODUCTION

The mechanisms of fragmentation which can operate in a wide variety of molecular ions have been discussed in detail in Chapter 3. To facilitate the use of these fragmentation mechanisms in structural elucidation the following summaries are included in which the mechanisms applicable to each particular type of functional group are outlined together. Spectra of known molecules are shown at the beginning of each summary with the prominent peaks labelled so that practice can be gained in applying the correct mechanism. In addition, unknown spectra included at the end of each summary provide practice in interpretation. To further simplify the problem of analysing unknown spectra a general method has been included. Some clues, which will in many cases indicate the type of molecule under examination from the major peaks, are also included because when the type of molecule is recognized the particular mechanisms applicable to that type can be applied, thus avoiding a lot of trial and error. The mass spectrometric shift technique which is often useful in interpreting mass spectra is also discussed. To illustrate the general approach to the determination of the structure of a molecule from its mass spectrum, six simple, worked examples have been included. Sources of useful information about mass spectrometry are given at the end of the chapter.

SUMMARIES OF FRAGMENTATION PROCESSES APPLICABLE TO PARTICULAR TYPES OF COMPOUNDS

Aliphatic alcohol, thiol or primary amine

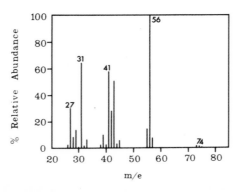

FIG. 1. Mass spectrum of n-butanol.

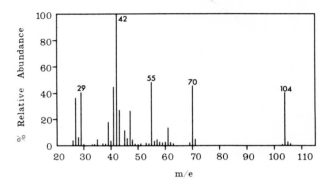

FIG. 2. Mass spectrum of n-pentanethiol.

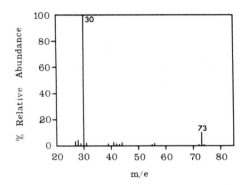

FIG. 3. Mass spectrum of n-butylamine.

Simple α-fission
See p. 40

$$R\overset{\cdot}{-}CR_2-\overline{X} \longrightarrow R\cdot\cdot\overset{\frown}{CR_2}\overset{\cdot}{-}\overset{+}{X} \longrightarrow CR_2\overset{+}{=}X$$

$CR_2\overset{+}{=}OH$ m/e **31** + n x 14
$CR_2\overset{+}{=}SH$ m/e **47** + n x 14
$CR_2\overset{+}{=}NH_2$ m/e **30** + n x 14

The value of *n* depends on the value of R which can be hydrogen or alkyl. The ion corresponding to the loss of the largest of the alkyl radicals is the most intense.

Rearrangement from the molecular ion (random)
See p. 53

This is significant for alcohols and mercaptans, but not for amines. The loss of H_2O and H_2S respectively occurs from the molecular ion to yield an 'olefine ion' of mass M − 18 and M − 34 respectively. These losses also occur thermally to yield an olefine neutral molecule before ionization. The extent of this decreases in the order alcohol>thiol>amine. There

$$\left[\begin{array}{c} CH_2-X \\ (CH_2)n \\ CH_2-H \end{array} \right]^{\ddot{+}} \longrightarrow \left[\begin{array}{c} CH_2 \\ (CH_2)n \\ CH_2 \end{array} \right]^{\ddot{+}} + HX$$

is, therefore, a similarity to the spectra of the corresponding olefines, being greatest with alcohols and least with amines.

Rearrangement from the fragment ion
See p. 58

If in the ion $CR_2=\overset{+}{X}$ one group R is ethyl or greater an olefine molecule can be eliminated. A four-membered transition state mechanism can be drawn:

$$R'-\underset{}{CH}-CH_2-\underset{\underset{+}{X}}{\overset{\overset{H}{|}}{\underset{\|}{C}}}-R \longrightarrow R'-\underset{H}{\overset{|}{C}}=CH_2 + H-\underset{\underset{+}{X}}{\overset{\|}{C}}-R$$

The three spectra at the beginning of the section have the significant ions labelled so that these can be correlated with the processes outlined. Below three spectra are included to give practice in the interpretation of unknown spectra.

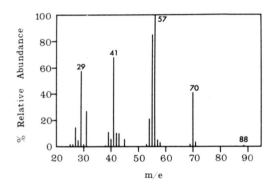

Unknown spectrum 1.

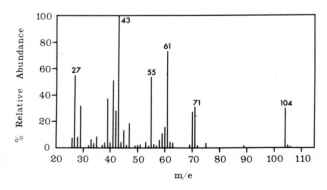

Unknown spectrum 2.

INTERPRETATION OF THE MASS SPECTRUM

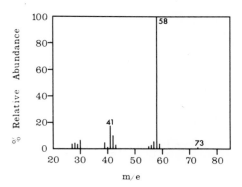

Unknown spectrum 3.

Aliphatic ether, thioether, secondary and tertiary amine

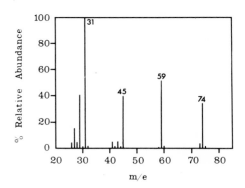

FIG. 4. Mass spectrum of diethyl ether.

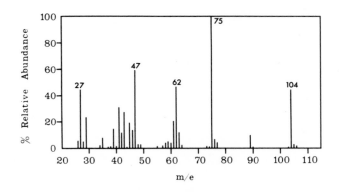

FIG. 5. Mass spectrum of ethyl n-propyl thioether.

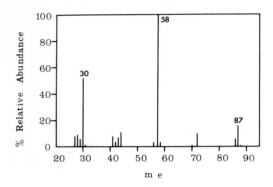

FIG. 6. Mass spectrum of ethyl n-propyl amine.

Simple α-fission
See p. 48

$$R \cdots CR_2 - \overset{+}{X} - \underset{R}{\overset{R}{C}} - R' \longrightarrow CR_2 = \overset{+}{X} - \underset{R}{\overset{R}{C}} - R'$$

$$CR_2 = \overset{+}{O} - \underset{R}{\overset{R}{C}} - R' \qquad m/e \; 45 + n \times 14$$

$$CR_2 = \overset{+}{S} - \underset{R}{\overset{R}{C}} - R' \qquad m/e \; 61 + n \times 14$$

$$CR_2 = \overset{+}{N} - \underset{R}{\overset{R}{C}} - R' \qquad m/e \; 44 + n \times 14$$
$$\phantom{CR_2 = \overset{+}{N}}\underset{R''}{}$$

Again R can be hydrogen or alkyl, the α-fission occurring in the longest chain preferentially. The value of *n* depends on the value of R.

Rearrangement from the fragment ion
See p. 58

When one of the groups in the fragment ion, attached to the carbon atom singly bonded to the heteroatom, and shown as R', is methyl or greater an olefine molecule can be eliminated. Consider the case of R' = CH$_3$:

$$CR_2 = \overset{+}{X} - CR_2 - \overset{H}{C}H_2 \longrightarrow CR_2 = \overset{+}{X}H$$

$$CR_2 = \overset{+}{O}H \qquad m/e \; 31 + n \times 14$$
$$CR_2 = \overset{+}{S}H \qquad m/e \; 47 + n \times 14$$
$$CR_2 = \overset{+}{N}HR'' \qquad m/e \; 30 + n \times 14$$

With tertiary amines an additional group R″ is present, which can allow this elimination to occur again if it is ethyl or greater. Rearrangement is also possible in fragment ions of this kind if R′ is ethyl or greater, e.g. R′ = ethyl:

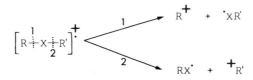

Simple fission of the carbon–heteroatom bond
See p. 43

This is significant in ether and thioether spectra, but not in amine spectra. When this fission process occurs in the molecular ion of an ether the charge remains with the alkyl

$$\left[R \!-\! X \!-\! R' \right]^{+\bullet} \begin{array}{c} \nearrow^1 \quad R^+ \; + \; {}^\bullet XR' \\ \searrow_2 \quad RX^\bullet \; + \; {}^+R' \end{array}$$

fragment and the structures of R and R′ determine whether fission (1) or (2) will be preferred. In thioethers with long alkyl chains the charge can remain with the sulphur-containing fragment. This behaviour is ascribed to the ability of sulphur to accommodate the positive charge more efficiently than oxygen, enabling it to form a cyclic ion as follows:

$$\left[C_5H_{11}\!-\!S\!-\!R \right]^{+\bullet} \longrightarrow \left[C_5H_{11}S \right]^{+} \longrightarrow \langle\!+\!SH$$

Unknown spectra are included below:

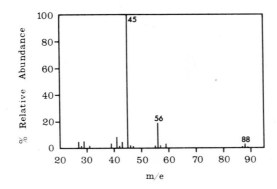

Unknown spectrum 4.

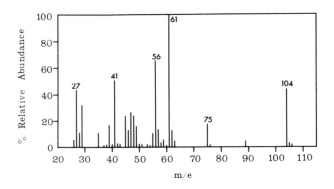

Unknown spectrum 5.

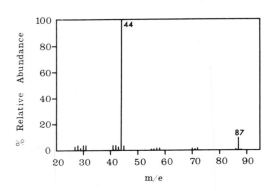

Unknown spectrum 6.

Aliphatic compounds containing a carbonyl group or the sulphur-containing analogue

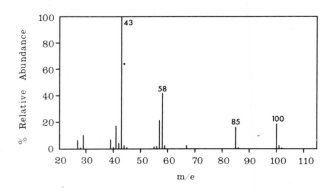

FIG. 7. Mass spectrum of methyl isobutyl ketone.

INTERPRETATION OF THE MASS SPECTRUM

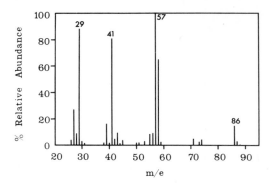

FIG. 8. Mass spectrum of 2-methylbutanal.

Simple fission of the bond α to the C=X group
See p. 44

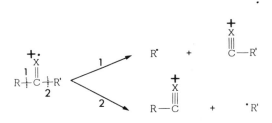

Fission in the fragment ion
See p. 48

The neutral molecule CO can be eliminated from the fragment ions containing it.

$$R-C\equiv O^+ \longrightarrow R^+ + CO \qquad R'-C\equiv O^+ \longrightarrow R'^+ + CO$$

McLafferty rearrangement
See p. 50

With aldehydes this can only occur as shown because R′ is hydrogen. With ketones R′ can be n-propyl or greater, and can present an alternative path. R″ can present another alternative path if it is ethyl or greater. With aldehydes it is also possible for a McLafferty rearrangement to occur so that the positive charge remains with the olefine fragment. The mechanism is shown below:

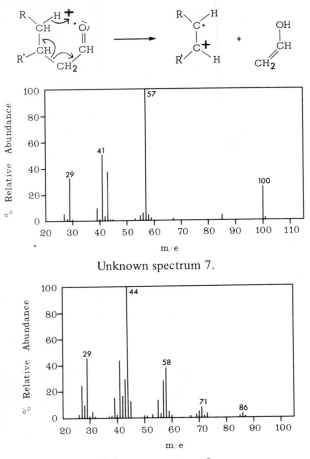

Unknown spectrum 7.

Unknown spectrum 8.

Carboxylic acid derivatives

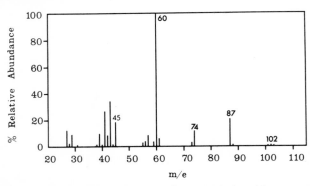

FIG. 9. Mass spectrum of n-pentanoic acid.

INTERPRETATION OF THE MASS SPECTRUM 79

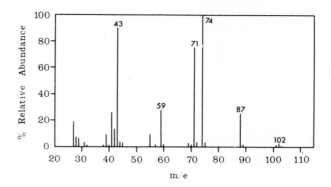

FIG. 10. Mass spectrum of methyl n-butanoate.

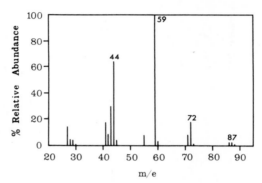

FIG. 11. Mass spectrum of n-butanamide.

Simple fission of the bond α to the C=X group
See p. 44

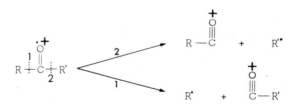

Process (1) is favoured in esters and (2) in acids and amides.

Fragment ion fission
See p. 48

R—C≡O⁺ formed from esters can lose CO, i.e.

$$R—C\equiv\overset{+}{O} \longrightarrow R^+ + CO$$

McLafferty rearrangement
See p. 50

This occurs in acids, amides and esters involving atoms in which R must be at least n-propyl. If a substituent R" is present as shown and is ethyl or greater it presents an

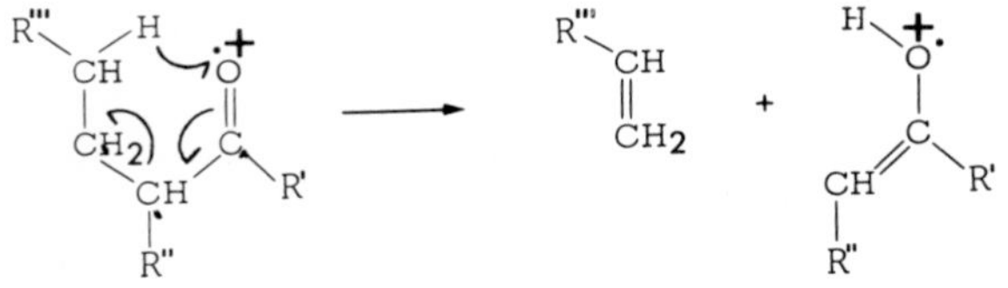

$$R'''\text{-CH}-CH_2-CH(R'')-C(=O^{+\cdot})-R' \; \xrightarrow{H} \; R'''\text{-CH}=CH_2 \;+\; H\text{-O}^{+\cdot}=C(R')\text{-CH}(R'')$$

alternative path. With esters, when R' is greater than O-ethyl another alternative path is presented. In favourable cases the aryl or alkyl portion in R' can compete for the positive charge. The mechanism is envisaged as:

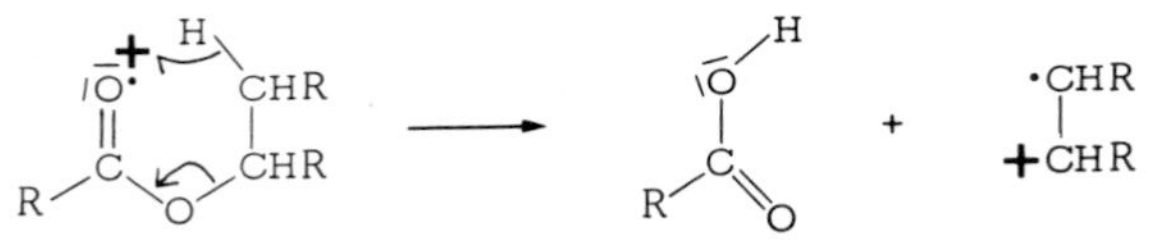

$$R\text{-C}(=O)\text{-O-CHR-CHR} \;\xrightarrow{H}\; R\text{-C}(\text{-O}^{\cdot})=O\text{-H} \;+\; {}^{\cdot}\text{CHR-}{}^{+}\text{CHR}$$

Random elimination of the acid part of an ester can also occur. Any of the ions formed by a McLafferty rearrangement can repeat the process if they fulfil the necessary structural requirements.

Double rearrangement
See p. 54

In esters where R' is greater than O-methyl a double rearrangement can occur whereby two hydrogen atoms are abstracted from the neutral fragment being eliminated. A possible mechanism is:

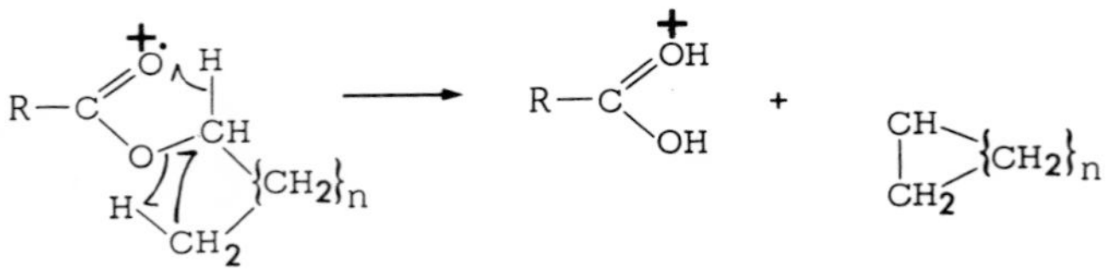

$$R\text{-C}(=O^{+\cdot})\text{-O-CH(CH}_2)_n\text{-CH}_2\text{-H} \;\longrightarrow\; R\text{-C}(\text{OH})_2^{+} \;+\; CH=CH\text{-(CH}_2)_n$$

In secondary and tertiary amides, fragmentation processes typical of amines can occur if the substituents on the nitrogen atom are of suitable structure.

Unknown spectra are shown at the top of the next page:

INTERPRETATION OF THE MASS SPECTRUM 81

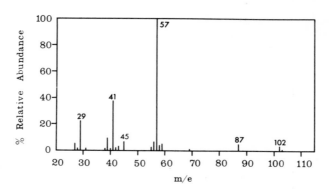

Unknown spectrum 9.

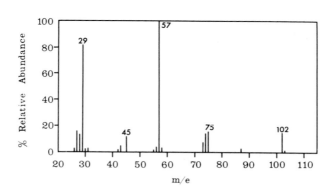

Unknown spectrum 10.

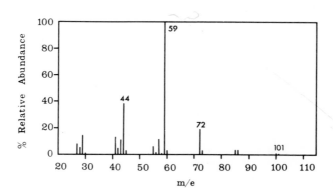

Unknown spectrum 11.

Cyclic systems with the functional groups carbonyl, alcohol, ether, thiol, thioether, ethylene ketal, ethylene thioketal and amine (all classes)

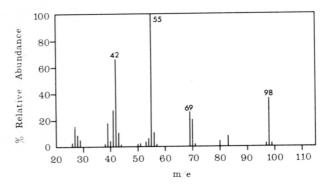

FIG. 12. Mass spectrum of cyclohexanone.

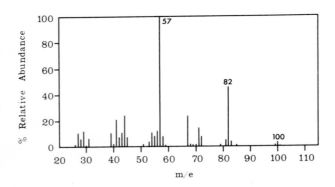

FIG. 13. Mass spectrum of cyclohexanol.

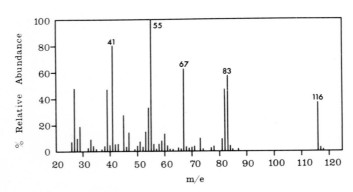

FIG. 14. Mass spectrum of cyclohexanethiol.

INTERPRETATION OF THE MASS SPECTRUM

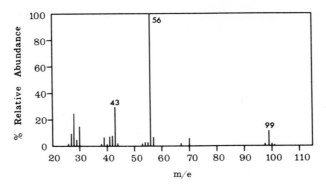

FIG. 15. Mass spectrum of cyclohexylamine.

Random elimination
See p. 53

Loss of H_2O and H_2S can occur from the alcohol and thiol respectively. In cyclic ketones an extensive rearrangement can also lead to loss of H_2O.

Complex fission
See p. 62

A   B   C   B

The ions shown below are formed when this occurs with the following functional groups: carbonyl, alcohol, ether, thiol, thioether, ethylene ketal, ethylene thioketal and amine respectively. If the cyclic system was suitably substituted the masses of these ions would

| m/e | 55 | 57 | 56 + R | 73 | 72 + R | 99 | 131 | 54+R_1+R_2 |

increase correspondingly for alkyl groups by $n \times 14$. Loss of alkyl fragments from ion C can occur by single electron transfers. Various processes have been postulated for formation of alkyl ions, the two following being representative for a cyclic ketone:

$$\overset{+}{C}H_2-CH=CH_2 \longleftrightarrow CH_2=CH-CH_2^+$$

84 INTERPRETATION OF THE MASS SPECTRUM

Unknown spectra are shown below:

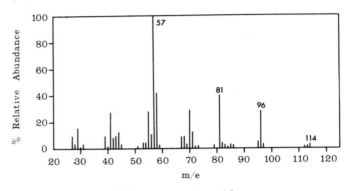

Unknown spectrum 12.

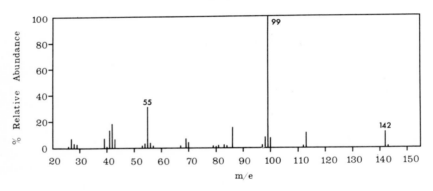

Unknown spectrum 13.

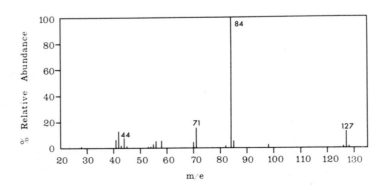

Unknown spectrum 14.

INTERPRETATION OF THE MASS SPECTRUM   85

Halogen-containing compounds

These show the characteristic relative abundance ratios of the peaks M, M + 2, M + 4 etc. depending on the halogen substituents present as described in the section on isotopes in Chapter 2.

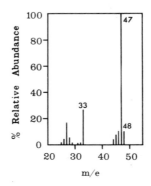

FIG. 16. Mass spectrum of fluoroethane.

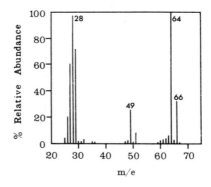

FIG. 17. Mass spectrum of chloroethane.

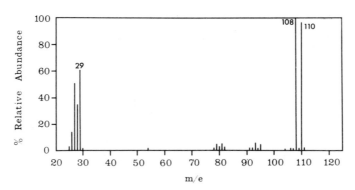

FIG. 18. Mass spectrum of bromoethane.

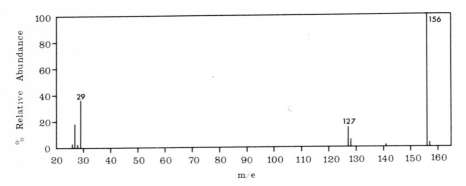

FIG. 19. Mass spectrum of iodoethane.

Fission of carbon–halogen bond
See p. 43

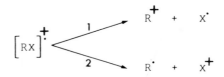

With fluorides and chlorides process (1) is favoured and with bromides and iodides process (2) is favoured.

Fission of the α-carbon–carbon bond
See p. 40

$$R-CH_2-\overset{+\cdot}{X} \longrightarrow CH_2=\overset{+}{X} + R^\cdot$$

$$R-CH_2-X \longrightarrow R^+ + CH_2=X^\cdot$$

This mechanism is favoured in the decreasing order $F > Cl > Br > I$. With fluorides the ion corresponding to the elimination of the smallest alkyl radical is the most intense, the reverse being the case with the other halogens.

Loss of hydrogen halide
See p. 53

This is most favoured with fluorides and chlorides and occurs by the random process described at the beginning of the rearrangement section in Chapter 3. Loss of H_2X occurs in lower molecular weight iodides and bromides. Ions of the structure are important in the spectra of chlorides and bromides having more than five carbon atoms in the alkyl chain, and are a result of the stability of the cyclic onium ion formed.

Detail on further points can be obtained from the textbook of Djerassi.[6]

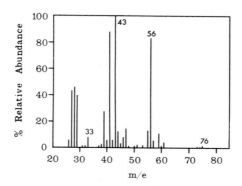

Unknown spectrum 15.

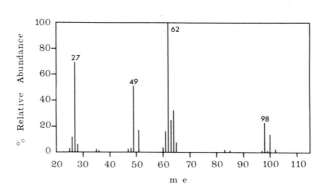

Unknown spectrum 16.

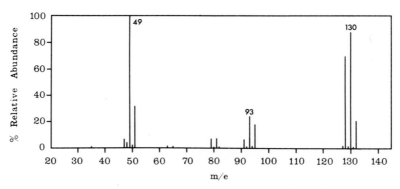

Unknown spectrum 17.

88 INTERPRETATION OF THE MASS SPECTRUM

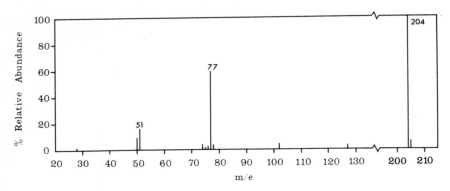

Unknown spectrum 18.

Outline of some fragmentation processes possible in a molecular ion containing an aromatic nucleus

Not all types of systems or fragmentation processes are represented in this summary, but the general type of mechanism is illustrated, and should enable the reader to approach new cases by analogy. Most of the following processes have been discussed earlier. See p. 46 regarding the way ions from aromatic compounds are depicted in this book.

1. Fission (labelled A) of the bond in the substituent group β to the aromatic system, e.g.

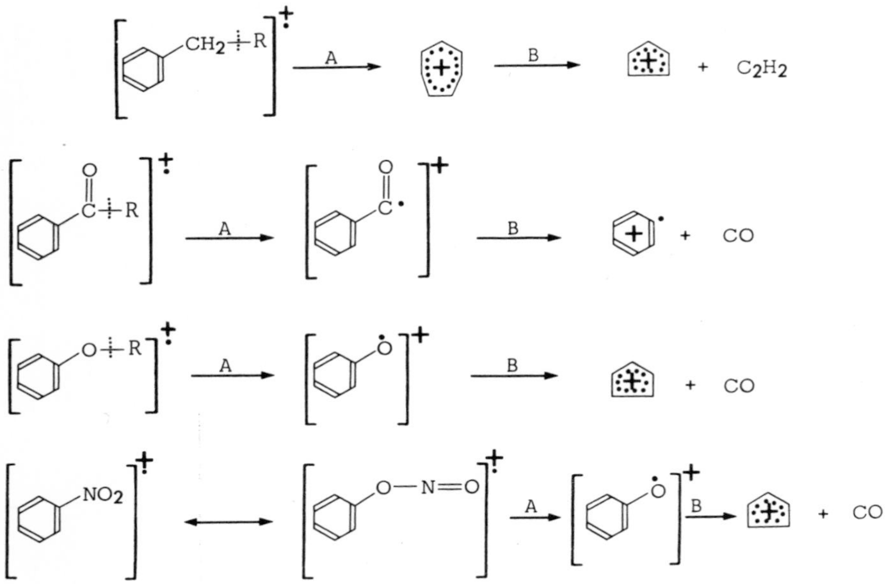

The structure of R affects the extent to which this fission occurs.

2. The ion formed by the simple fission process outlined above eliminates a stable neutral molecule. This has been indicated in the above figures by B, but the exact mechanism is not clear except in the case of C_6H_5CO which was mentioned earlier.

3. Simple fission of the bond from the substituent group to the aromatic system can also occur with the charge being preferentially retained on the aromatic system, e.g.

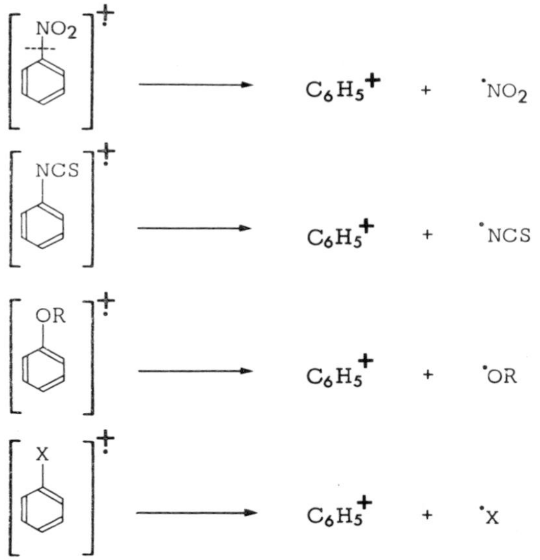

$$[C_6H_5NO_2]^{+\cdot} \longrightarrow C_6H_5^+ + \cdot NO_2$$

$$[C_6H_5NCS]^{+\cdot} \longrightarrow C_6H_5^+ + \cdot NCS$$

$$[C_6H_5OR]^{+\cdot} \longrightarrow C_6H_5^+ + \cdot OR$$

$$[C_6H_5X]^{+\cdot} \longrightarrow C_6H_5^+ + \cdot X$$

where X is halogen. The neutral fragment should have reasonable stability for this fission to occur, but in many cases the aromatic system is of very high stability and there is no favourable alternative.

4. An apparent McLafferty rearrangement utilizing one of the double bonds of the aromatic system can occur if the conditions are fulfilled by the substituent, e.g.

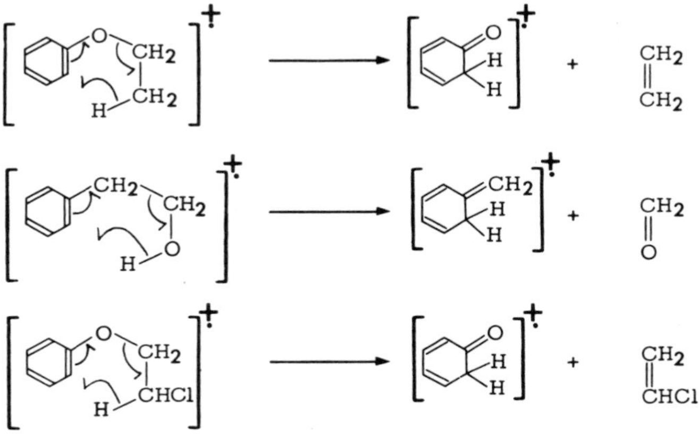

5. Some compounds can eliminate a neutral molecule possibly by means of a four-membered transition state as shown over page:

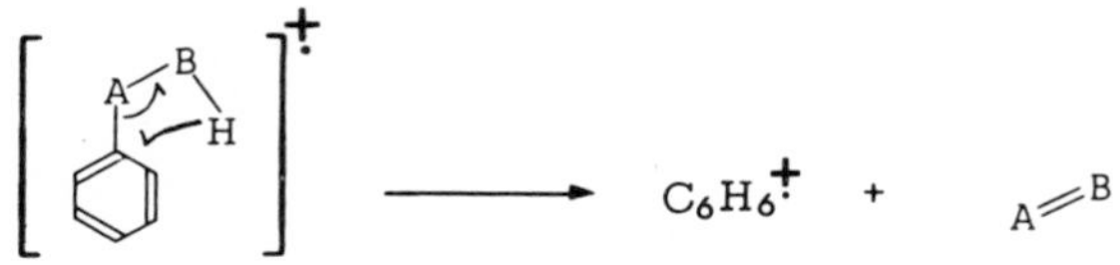
A
B
H
C6H6
+
A=B

For example:

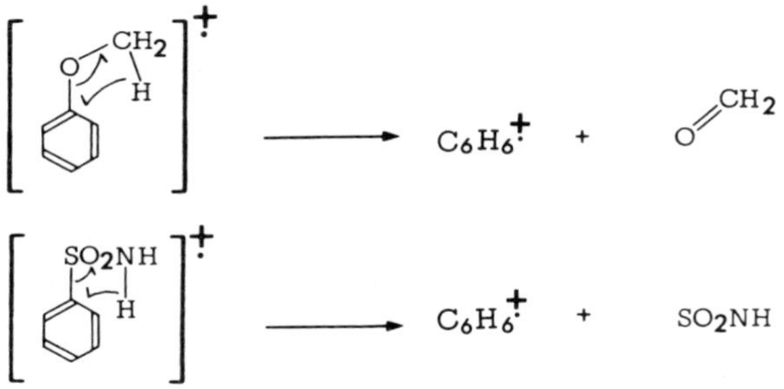
O
CH2
H
C6H6
+
CH2
SO2NH
H
C6H6
+
SO2NH

6. *Ortho*-disubstituted aromatic systems can eliminate a neutral molecule by rearrangement if the functional groups are suitable. This provides a means of differentiating an

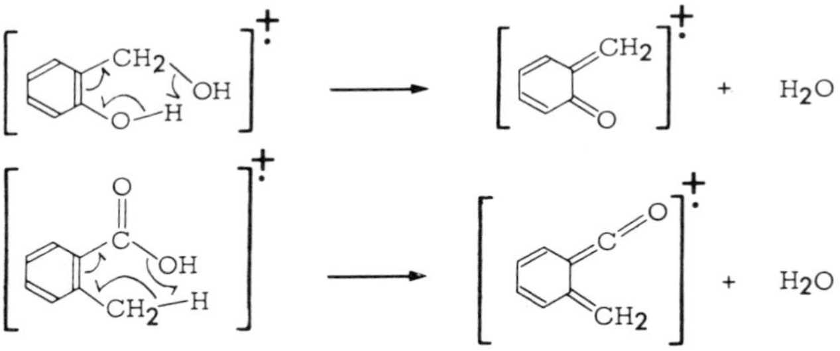
CH2
OH
O
H
CH2
O
+
H2O
O
C
OH
CH2
H
C=O
CH2
+
H2O

ortho-disubstituted molecule from its *meta*- and *para*-isomers.

7. When a saturated ring is attached to an aromatic system a retro-Diels–Alder process can occur utilizing one of the double bonds in the ring.

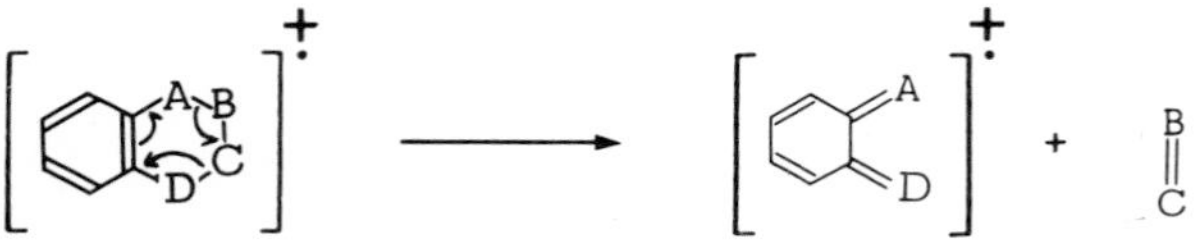
A
B
C
D
A
D
+
B
C

For example:

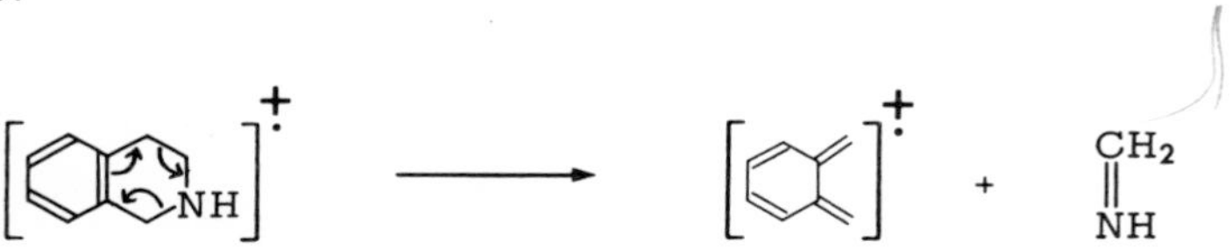
NH
+
CH2
NH

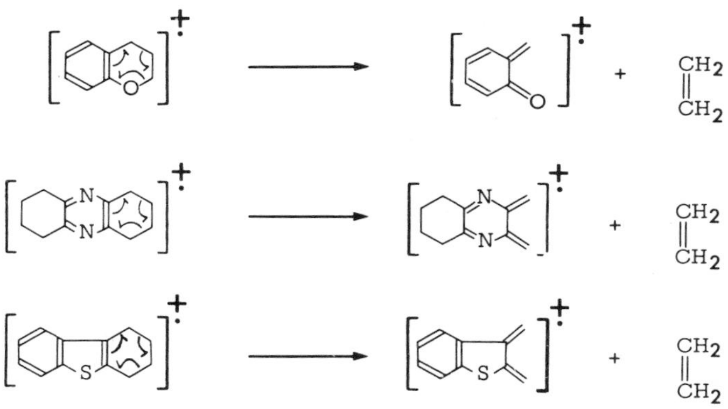

[Diagrams of aromatic fragmentation reactions showing loss of CH₂=CH₂ from benzofuran, quinoxaline-type, and benzothiophene cation radicals]

8. Aromatic nuclei which include carbonyl groups can eliminate these as neutral molecules as follows:

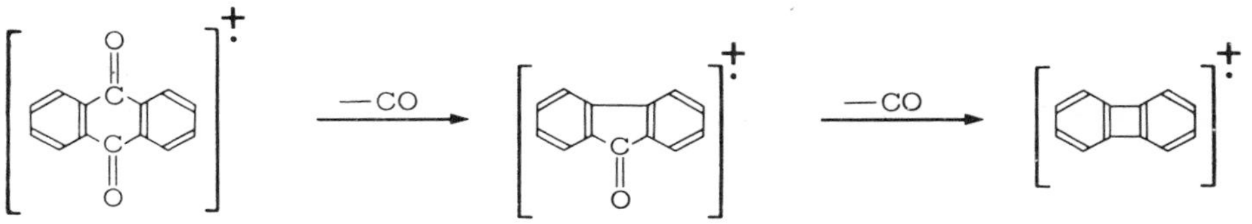

[Diagram: anthraquinone cation → −CO → fluorenone cation → −CO → biphenylene cation]

In the case of a phenolic hydroxyl group a four-membered transition state enables the carbonyl group to be formed. It is subsequently eliminated as above, i.e.

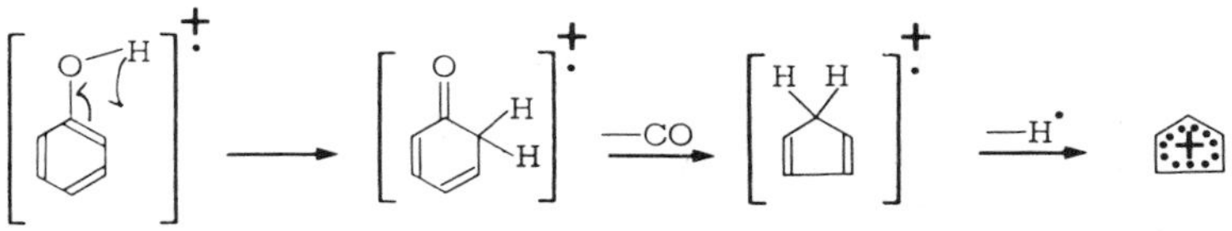

[Diagram: phenol cation → cyclohexadienone cation → −CO → cyclopentadiene cation → −H• → cyclopentadienyl cation]

This also occurs when the substituent is NH_2; the neutral molecule eliminated is HCN. When several functional groups are present in a molecule containing an aromatic nucleus or when several fragmentation processes are possible with one functional group an interplay of stability factors related to the arrangement of atoms in the molecule will determine which of the processes 1–8 will occur, and to what extent the ions formed will dominate the mass spectrum. Analysis of the spectrum is a trial and error process and can be time-consuming.

AN APPROACH TO THE DETERMINATION OF THE STRUCTURE OF A MOLECULE WITH THE AID OF MASS SPECTROMETRY

This approach is based on the availability of a low resolution mass spectrum only, the extra advantages of a knowledge of the ionic formulae being indicated where appropriate. It should be emphasized that it is very difficult to determine the structure of a molecule with no more information than its mass spectrum available. In the process of interpretation of a mass spectrum any other information available should be used as an aid and to check any conclusions made. The scheme given over page is only a general outline; it can be modified as appropriate.

1. Identify the m/e values of the ions in the spectrum (see p. 10).
2. Try to locate the molecular ion (see p. 18). Remember that in some cases the molecular ion may not be observed. It is also possible that the most intense peak in the highest m/e group may correspond to a $[M-H]^+$, $[M-H_2]^+$ or $[M+H]^+$ ion. The intensity of the last of these, relative to the others, is pressure-dependent and so variation of the sample pressure will prove or disprove this point.
3. Having provisionally identified the molecular ion the relative abundances of the ions in its m/e region should be checked to see if any elements with isotopes (abundant) are present. Such elements include chlorine, bromine, silicon, sulphur, boron and many of the metals. Any qualitative analytical data showing the presence of any other elements should also be taken into account at this stage. If the accurate ionic weight is known, then the molecular formula can now be deduced (see p. 23). If quantitative analytical data are available, the unit m/e value of the molecular ion should be checked for consistency with these data. If the molecular formula has been determined by accurate mass measurement, this should also be checked for consistency. If the two sets of data are inconsistent, then it is likely that the assignment of the molecular ion peak is incorrect. If this is so or the analytical data are not available, the deductions should be checked by running the spectrum of an appropriate derivative (see p. 33) and repeating the above procedure.
4. If the molecular formula has been found by mass measurement check the number of double bond equivalents present i.e. rings or double bonds. The method is as follows: C_nH_{2n+2} is fully saturated. When heteroatoms are present in a molecule these are substituted by carbon and hydrogen to get a formula in terms of carbon and hydrogen. This is subtracted from the fully saturated formula with the same number of carbon atoms and the difference which represents a number of hydrogen atoms is divided by two to yield the number of double bond equivalents. Nitrogen is substituted by CH, oxygen and sulphur by CH_2, and halogens by CH_3, e.g. if a molecule had a formula $C_7H_{11}N_2OSCl$, substitution for the heteroatoms would yield $C_7H_{11}(CH)_2(CH_2)CH_2 \cdot (CH_3) = C_{12}H_{20}$. The C_{12} saturated molecule is $C_{12}H_{26}$. Subtraction yields six and the number of double bond equivalents is three.
5. The m/e values of the (prominent) ions should be noted and their formulae, if possible, evaluated using the relevant points in section 3.
6. Note the intensity of the molecular ion peak. This will indicate the stability of the molecular ion and will sometimes indicate the class of compound. Thus an intense molecular ion is often associated with an aromatic compound. The presence of doubly charged ions is also sometimes indicative of this.
7. Find the mass differences between the molecular ion and the fragment ions and between the fragment ions themselves. When exact mass measurements have been carried out, the formulae of the (prominent) ions are known and so the formulae of the neutral fragments lost can be determined by subtraction.
8. Try to identify fragmentation sequences using the metastable ions (see p. 19). The mass difference between two peaks which are apparently connected by a metastable must be chemically reasonable, and the ionic formulae, when available, can be used to check this. Thus, the daughter ion cannot have more atoms of any element than the parent ion and the loss of fragments like NH, CH_6 etc. is unlikely. If the origin of a prominent ion is unknown, where possible, a metastable can be sought (see p. 22) in the first field-free region.

9. From the above evidence and any other information available consider structures or partial structures. The possibility of the formation of some of the ions by the fragmentation reactions outlined in Chapters 3 and 4 should be investigated. To help in this it should be remembered that the difference between processes in which molecules are expelled and those in which radicals are expelled is that, unless the fragment lost contains an odd number of nitrogen atoms, then loss of an even mass fragment corresponds to the loss of a molecule and *vice versa*. If the fragment lost contains an odd number of nitrogens, then the reverse is the case. In the case of a molecular ion or an odd-electron ion the loss of a neutral molecule implies a rearrangement process.
10. If a structure has been deduced from some of the available evidence then it must be consistent with the formation of all the reasonably intense ions in the spectrum. If a reference spectrum is available the comparison of the two spectra should prove or disprove the proposed structure. Unlike many other physical analytical techniques, if the two spectra were not run on the same machine under approximately the same conditions they may differ not only in the relative intensities of various ions, but in some cases one spectrum may contain new ions due to pyrolysis. In this case, for a new compound, and also where there is the possibility of geometrical isomerism, the structure must be confirmed by other means.

CHARACTERISTIC FRAGMENTATION PATTERNS

In this section the characteristic fragmentation of some of the simple compounds discussed in this book is summarized. This will be of help in trying to decide which class of compound an unknown spectrum represents.

Saturated aliphatic hydrocarbon

A homologous series of pairs of peaks is observed corresponding to ions of the formula C_nH_{2n-1} and C_nH_{2n+1}. The maximum abundance is reached when n is 3, 4 or 5 and falls off as n increases. Points of chain branching are indicated by peaks in the homologous series being more abundant than their neighbours $n + 1$ and $n - 1$.

Olefinic hydrocarbon

Very similar to saturated hydrocarbon spectra, except that the series of peaks is shifted two mass units down scale. Abundant even mass peaks corresponding to rearrangement ions formed by processes involving the double bond occur with m/e values in the homologous series $28 + n \times 14$, depending on the chain length and the position of the double bond.

Aliphatic alcohol

$[M - H_2O]^+$ ion. Peak or peaks in the homologous series m/e $31 + n \times 14$ depending on structure. Strong resemblance to the spectrum of the corresponding olefine formed by elimination of water.

Aliphatic thiol

$[M - H_2S]^+$ ion. Peak or peaks in the homologous series m/e $47 + n \times 14$ depending on structure. Resemblance to the spectrum of the corresponding olefine formed by elimination of H_2S. The isotope peak at m/e $M + 2$ will be 4 per cent of the intensity at m/e M due to the presence of ^{34}S. Peaks can also occur at m/e 32, 33 and 34, corresponding to $^{32}S^+$, $H^{32}S^+$ and $H_2^{32}S^+$.

Aliphatic amine
Odd mass molecular ion peak. α-fission peaks and the peaks due to fragment ion rearrangement processes occur in the homologous series m/e $30 + n \times 14$.

Aliphatic ether
α-fission ions and the peaks due to fragment ion rearrangement processes occur in the homologous series m/e $31 + n \times 14$. $[M - OR]^+$ ions also occur in the homologous series m/e $29 + n \times 14$.

Aliphatic thioether
Peak at m/e $M + 2$ due to the ^{34}S isotope is 4 per cent of the abundance of M. The homologous series of peaks analogous to those in the ether spectrum formed by α-fission and rearrangement, is m/e $61 + n \times 14$.

Aliphatic aldehyde
$[M - 1]^+$ ion owing to loss of aldehydic hydrogen. α-fission ion at m/e 29. Peaks corresponding to alkyl fragments in series m/e $29 + n \times 14$. Peaks at $M - (44 + n \times 14)$ and peaks at m/e $44 + n \times 14$ due to rearrangement. n refers to substituents on the carbon atom α to the carbonyl group.

Aliphatic ketone
Peaks due to $[M - R_1]^+$ and $[M - R_2]^+$ ions and also R_1^+ and R_2^+ in homologous series m/e $43 + n \times 14$ and m/e $29 + n \times 14$ ($CO = C_2H_4$ in mass). Loss of olefine molecule by McLafferty rearrangement yields peak or peaks in the series m/e $58 + n \times 14$, depending on the structure.

Aliphatic carboxylic acid
$[M - CO_2H]^+$ ion and peak at m/e 45. Rearrangement peak at $M - (46 + n \times 14)$ or in series m/e $46 + n \times 14$ depending on structure. Also peaks in the series m/e $29 + n \times 14$ due to alkyl fragments.

Aliphatic ester
$[M - OR]^+$ ions in series m/e $29 + n \times 14$. Rearrangement peak corresponding to an ion containing the carboxyl group in the series m/e $46 + n \times 14$, or due to loss of the acid moiety in the series $M - (46 + n \times 14)$. With esters greater than methyl double rearrangement is possible to yield a peak in the homologous series m/e $47 + n \times 14$. This peak identifies the acid part of the ester, e.g. butanoates show m/e 89 peak. The McLafferty rearrangement is favoured in the acid part if the choice exists. The generated double bond then allows the processes to occur in the alcohol side.

Aliphatic amide
Odd mass molecular ion peak. Even mass peak in the series m/e $44 + n \times 14$ due to simple fission (α to carbonyl and β to the nitrogen atom). Rearrangement peak of odd mass in the homologous series m/e $59 + n \times 14$. In secondary and tertiary amides, amine-type β-fission followed by fragment ion rearrangement can occur to yield ions of m/e $30 + n \times 14$.

Aliphatic fluoride
Peak at m/e 19 corresponding to F^+. Peaks at m/e M − 19 and M − 20 (due to loss of F and HF respectively). Smallest R group is lost by α-fission.

Aliphatic chloride
Characteristic abundance ratios of peaks at m/e M, M + 2, M + 4 etc. Peaks at m/e 35 and 37 in ratio of 3:1 and at m/e 36 and 38 in the ratio of 3:1 corresponding to ^{35}Cl and ^{37}Cl and $H^{35}Cl^+$ and $H^{37}Cl^+$ respectively. M − 35 and M − 36 peaks are also observed due to loss of Cl and HCl from the molecular ion.

Aliphatic bromide
Analogous to chlorides except that the bromine isotope masses are 79 and 81 in abundance ratios of 1:1. Loss of H_2Br from M^+ also occurs.

Aliphatic iodide
Peak at 127 corresponding to I^+. Peaks due to loss of I and H_2I from molecular ion.

Aliphatic nitrile
Odd mass molecular ion peak. Very abundant peak due to rearrangement in the series m/e 41 + $n \times$ 14. M + 1 peak due to ion–molecule collision with hydrogen abstraction.

Alicyclic molecules
For alicyclic molecules in the above classes of compounds the summary on p. 62 should be consulted. Many of the clues are directly applicable, e.g. the presence of $[M − H_2O]^+$, $[M − H_2S]^+$, $[M − \text{halogen}]^+$, the Nitrogen Rule and characteristic isotope peaks. The fact that at least one extra double bond equivalent is present brings this type of structure (or a structure containing an olefinic double bond) under consideration.

Aromatic compounds or substituted aromatic nuclei
The best clues to the presence of an aromatic system are:
1. The number of double bond equivalents.
2. The high percentage abundance of the peaks of high m/e in the mass spectrum (roughly those $> m/e$ 77 $[C_6H_5]^+$) and the sparsity of these peaks. For simple aromatic compounds the molecular ion peak is often the base peak or the base peak is represented by an ion of high m/e formed from the molecular ion by one simple fission process or one rearrangement process.
3. The presence of doubly charged molecular ion peaks occurring at $m/2e$ in the spectrum.
4. Due to the benzene ring fragmentation peaks usually occur at m/e 39, 50, 51, 52 and 65. A survey of most of the common fragmentation processes occurring in this type of compound has been given (see p. 46).

Two examples of the fundamental features are given below.

Alkyl benzenes
Base peak is either m/e 91 + $n \times$ 14 due to α-fission or m/e 92 + $n \times$ 14 due to rearrangement.

Aromatic aldehydes
M^+, $[M − 1]^+$ and $[M − HCO]^+$ are the significant ions.

THE MASS SPECTROMETRIC SHIFT TECHNIQUE

In cases where the structure of an unknown molecule has been partly determined, whether from its mass spectrum or otherwise, comparison with the spectrum of a compound having a known related structure can be used to give further structural information. The technique was given its name and formally outlined by Biemann,[5] but it is often used unconsciously. It is particularly useful if the problem is the location of some small groups on a known skeleton.

The principle is that, compared with the spectrum of the known compound, those ions in the spectrum of the unknown compound corresponding to portions of the unknown molecule which have the same structure as the reference compound will appear at the same m/e value. Those portions of the unknown molecule which contain structural modifications will appear at new m/e values compared with the reference compound and thus the mass and formula of the substituent can be found. This argument only applies when the structural modifications are such that they themselves do not undergo appreciable fragmentation, nor do they initiate any new or alter any old fragmentation routes.

Biemann used this technique in the case of alkaloids as indicated in the following example. From the spectra of ibogamine, ibogaine and tabernanthine, which differ only in the nature of their aromatic substitution (see Fig. 20), it can be seen that the ions at m/e 136 and 149 do not contain the aromatic portion of the molecule, since they occur at the same place in the three spectra. Since the ions at m/e 156 and 195 in the spectrum of ibogamine are replaced by ions at m/e 186 and 225 in the spectra of the other two compounds, these ions must contain the aromatic portion of the molecule. Thus from the spectrum of the fourth alkaloid, ibogaline, which has the same skeleton, it can be seen that its extra methoxy group (*cf.* tabernanthine and ibogaine) must be on the aromatic portion since the ions at m/e 186 and 225 due to this portion are replaced by ions at m/e 216 and 255, but the ions at m/e 136 and 149 are still present.

This technique is mainly of use with small substituents attached to a part of the skeleton which does not fragment itself, such as an aromatic ring. In principle, it is also possible to apply the technique to members of the same family substituted differently in the alicyclic part. The limitations imposed by the basic rules are, however, stringent because most of the important fragmentation processes are occurring in the alicyclic part. The peaks (or at least some of them) due to the aromatic part will, however, remain at the same m/e values when the substitution differences are occurring in the alicyclic part. In very favourable cases both the aromatic part and alicyclic part could be substituted and still enable the shift technique to be applied. The technique can be applied to any family of molecules which obeys the basic rules.

When experience has been gained in a field like alkaloids, for example, one becomes aware that characteristic peaks exist for different structural features and basic skeletons. It is therefore possible to select a known reference spectrum to compare with an unknown spectrum if any characteristic peaks are recognized. In the absence of a reference spectrum analysis of the mass spectrum in conjunction with other spectroscopic evidence and chemical information can yield the structure of the unknown.

EXAMPLES OF STRUCTURAL DETERMINATION USING MASS SPECTROMETRY

The following six examples illustrate the way in which the structure of a molecule is found using its mass spectrum. The molecular formula could easily be found using exact mass measurement and to simplify the elimination process discussed it is supplied in each case.

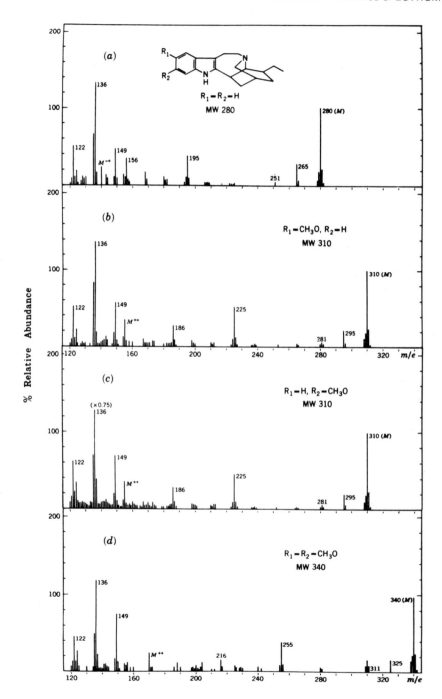

FIG. 20. Mass spectra of (a) ibogamine, (b) ibogaine, (c) tabernanthine, and (d) ibogaline. (Reproduced from *Mass Spectrometry* by K. Biemann by kind permission of McGraw-Hill. Copyright © 1962 by McGraw-Hill Book Co.).

Knowing the molecular formula is not necessary with simple compounds like those discussed because with a little experience the characteristic features of these mass spectra stand out and very little vetting is necessary. One should not set out to solve the structure of a complex molecule using only mass spectrometry unless it is not possible to use the other spectroscopic techniques. It will become clear with a little experience that some information becomes available from a mass spectrum for very little mental effort. This information, combined with the corresponding information from the other spectroscopic techniques, is often all that is required to confirm a structure absolutely. The further one tries to go with any single technique the more difficult and time-consuming the process becomes.

Example 1

There are no characteristic isotope peaks to suggest the presence of sulphur, chlorine, bromine etc. The molecular ion peak at m/e 88 makes the atomic constitutions $C_5H_{12}O$, $C_4H_{12}N_2$, $C_4H_8O_2$ possibilities. Using double bond equivalents the long list of possible compounds includes alcohols, ethers, diamines, acids, esters, hydroxy ketones, alkoxy ketones, hydroxy epoxides etc. Exact measurement shows the molecular formula to be $C_4H_8O_2$ which narrows down the number of possibilities. The fact that the base peak is at m/e 60, and must be formed by a rearrangement process eliminates all possibilities except acid and ester. The possibilities are:

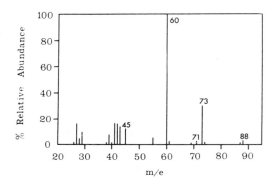

Only structures (3) and (5) could eliminate a neutral molecule of ethylene by means of a McLafferty rearrangement. If the structure was (3) there would be a peak of reasonable abundance at m/e 61 due to a double rearrangement process. This is not observed. The structure is, therefore, (5) and the compound is n-butanoic acid. The correlation is as follows:

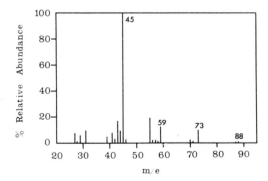

Simple fission is favoured α to the carbonyl group.

Example 2

There are no characteristic isotope peaks. The even-electron molecular ion could have a number of atomic constitutions $C_4H_8O_2$, $C_4H_{12}N_2$, $C_5H_{12}O$. There is a wide variety of possible compounds to be vetted using the clues. These include alcohols, ethers, acids, esters, diamines etc. To simplify the discussion exact mass measurement shows the molecular formula to be $C_5H_{12}O$. This molecule has no double bond equivalents and must, therefore, be an ether or an alcohol. An ether is ruled out because there is an M − 18 peak. The compound must, therefore, be an alcohol, and the possibilities are shown over page:

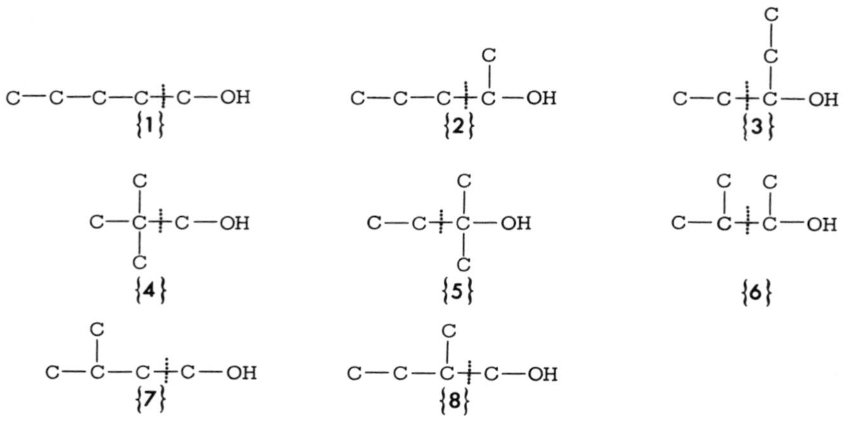

INTERPRETATION OF THE MASS SPECTRUM

```
C—C—C—C┼C—OH          C—C—C┼C—OH          C—C┼C—OH
       {1}                 |                    |
                           C                    C
                          {2}                   |
                                                C
                                               {3}

     C                     C                  C   C
     |                     |                  |   |
 C—C┼C—OH              C—C┼C—OH           C—C┼C—OH
     |                     |                
     C                     C                  
    {4}                   {5}                 {6}

     C                     C
     |                     |
 C—C—C┼C—OH            C—C—C┼C—OH
    {7}                   {8}
```

The base peak at m/e 45 is formed by simple fission of the α-C—C bond. This eliminates all the possibilities except (2) and (6). The ion of m/e 59 must contain the oxygen atom and is formed by a two-step process. The first step is loss of a hydrogen atom to yield an ion of m/e 87 which then eliminates a molecule of ethylene. By the allowed mechanisms this sequence can only occur from structure (2) involving a McLafferty rearrangement:

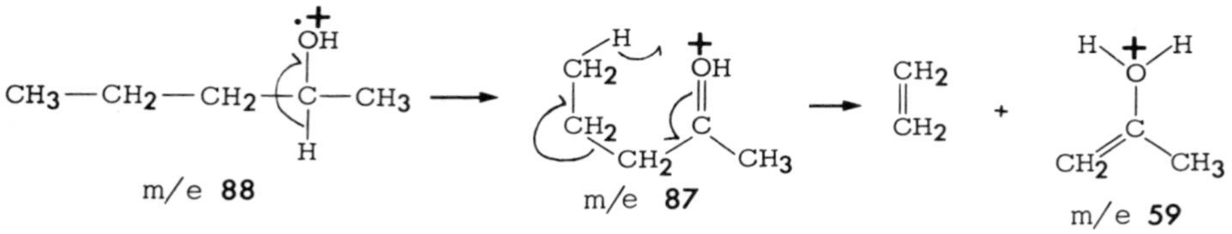

$$CH_3-CH_2-CH_2-\overset{\overset{\cdot+}{OH}}{\underset{H}{C}}-CH_3 \longrightarrow \text{[6-membered transition state]} \longrightarrow \underset{CH_2}{\overset{CH_2}{\|}} + \underset{CH_2}{\overset{H\diagdown\overset{+}{O}\diagup H}{\underset{\|}{C}}}_{CH_3}$$

m/e 88 → m/e 87 → m/e 59

Structure (6) which contains the isopropyl group does not enable a six-membered ring

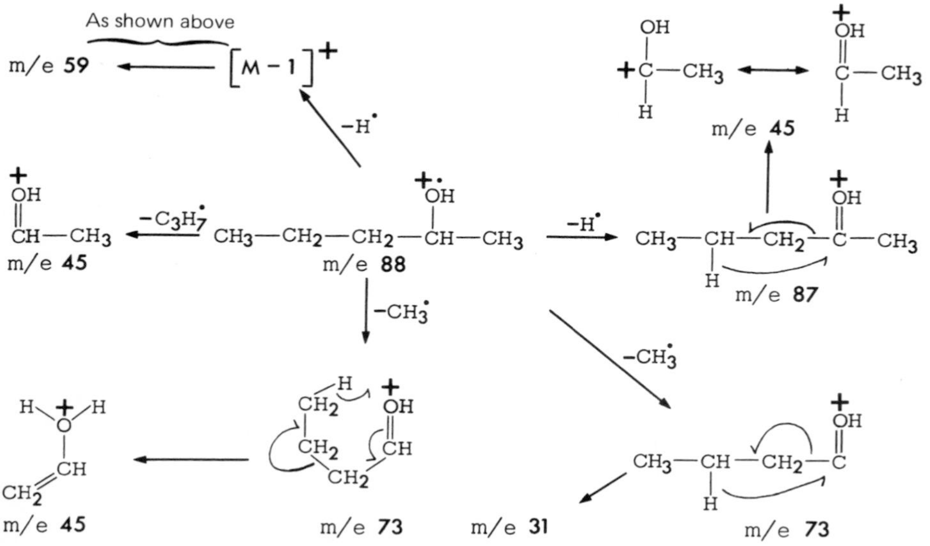

[Fragmentation scheme showing:]

m/e 59 ← As shown above ← [M − 1]$^+$

$\overset{+}{\overset{OH}{\|}}$CH—CH$_3$ (m/e 45) ← −C$_3$H$_7^{\cdot}$ — CH$_3$—CH$_2$—CH$_2$—$\overset{+\cdot}{\overset{OH}{|}}$CH—CH$_3$ (m/e 88) $\xrightarrow{-H^{\cdot}}$ CH$_3$—CH—CH$_2$—$\overset{+}{\overset{OH}{\|}}$C—CH$_3$ (m/e 87)

                                    OH $\overset{+}{\overset{OH}{\|}}$
                                    |
                                +C—CH$_3$ ←→   C—CH$_3$
                                    | |
                                    H   H
                                                         m/e 45

$-CH_3^{\cdot}$ ↓

H$\diagdown\overset{+}{O}\diagup$H $-CH_3^{\cdot}$
     $\|$
     CH   ←   [6-membered ring intermediate]   CH$_3$—CH—CH$_2$—$\overset{+}{\overset{OH}{\|}}$C
CH$_2$ m/e 73   m/e 31 |
m/e 45   H
                                                                              m/e 73

transition state to operate. Elimination of a neutral molecule of ethylene also occurs by the rearrangement shown on p. 100 from the ion m/e 73 formed from structure (2).

The elimination of a neutral olefine molecule from the fragment ion by means of a four-membered transition state (see p. 58) is not of diagnostic value in this case because both structures (2) and (6) would eliminate propylene. The structure, therefore, is (2) which is pentan-2-ol.

Example 3

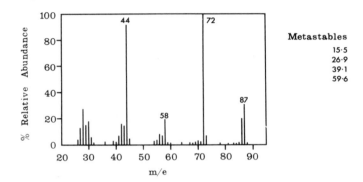

There are no characteristic isotope peaks. The odd mass molecular ion peak shows that the compound contains an odd number of nitrogen atoms. Without mass measurement the possible formulae are $C_5H_{13}N$, C_4H_9NO, $C_3H_9N_3$, $C_3H_5NO_2$, and these represent a myriad of possible compounds which would have to be vetted using the clues. By exact mass measurement the formula of the unknown compound is $C_5H_{13}N$. There are no double bond equivalents and so the compound must be an amine. The possible structures are as follows:

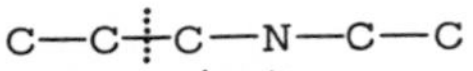

{10} C—C⫶C—N—C—C

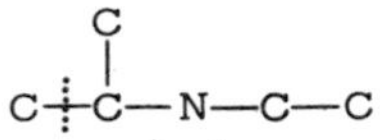

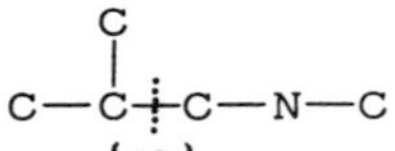

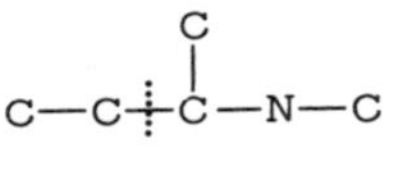

{11} C⫶C(—C)—N—C—C

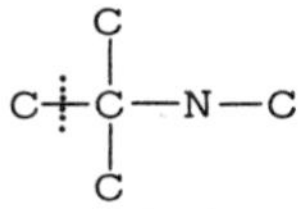

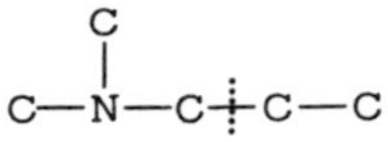

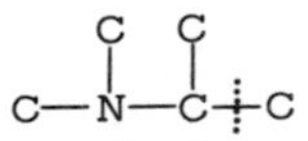

{12} C—C(—C)⫶C—N—C

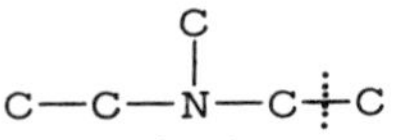

{13} C—C⫶C(—C)—N—C

{14} C⫶C(—C)(—C)—N—C

{15} C—N(—C)—C⫶C—C

{16} C—N(—C)(—C)—C⫶C

{17} C—C—N(—C)—C⫶C

The base peak at m/e 72 corresponds to the ion formed by loss of a methyl group from the molecular ion. The metastable peak at m/e 59.6 corresponds to the transition m/e 87 → m/e 72 + 15. Only structures (11), (14), (16) and (17) are suitably substituted for simple fission of the α-C–C bond (the favoured process) to result in loss of a methyl group. The metastable peak at m/e 26.9 shows that m/e 72 → m/e 44 + 28. Loss of a neutral molecule of ethylene can only occur from the $[M - CH_3]^+$ ions formed from structures (11) and (17). The second degradation sequence is loss of a hydrogen atom from the molecular ion followed by successive eliminations of two molecules of ethylene. The metastable peaks at m/e 39.1 and m/e 15.5 correspond to the transitions m/e 86 → m/e 58 + 28 and m/e 58 → m/e 30 + 28 respectively. This can only occur from structure (17). The unknown molecule is, therefore, methyl diethyl amine and the correlation is as follows:

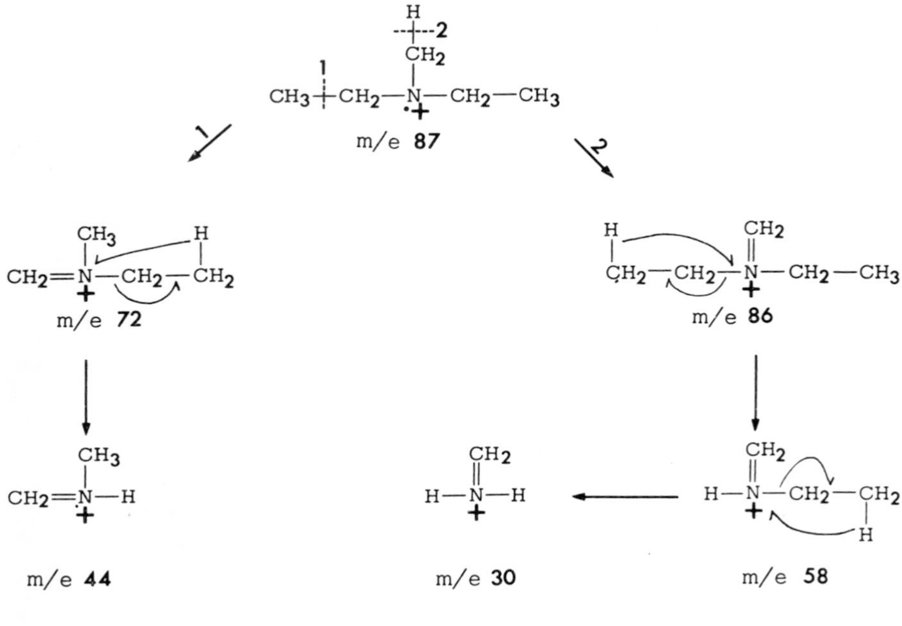

$$CH_3 \!\!\mid\!\! CH_2 - \overset{CH_2\text{-}H\,(2)}{\underset{+}{N}} - CH_2 - CH_3$$

m/e 87

Path 1:
$$CH_2 = \overset{CH_3}{\underset{+}{N}} - CH_2 - CH_2(H)$$
m/e 72

↓

$$CH_2 = \overset{CH_3}{\underset{+}{N}} - H$$
m/e 44

Path 2:
$$H-CH_2-CH_2-\overset{CH_2}{\underset{+}{N}}-CH_2-CH_3$$
m/e 86

↓

$$H-\overset{CH_2}{\underset{+}{N}}-CH_2-CH_2(H)$$
m/e 58

→

$$H-\overset{CH_2}{\underset{+}{N}}-H$$
m/e 30

Example 4

The molecular ion peak is the base peak and no characteristic isotope peaks are observed. The molecule is most probably aromatic. Of the possible molecular formulae only those having four or more double bond equivalents would, therefore, be considered. Exact mass measurement shows the molecular formula to be C_7H_8O. This unknown compound must have four double bond equivalents and the possible structures are alcohol, phenol or ether. The alcohol would be benzyl alcohol (1), the ether would be anisole (2) and the phenols would be o-, m- and p-cresol, (3), (4) and (5) respectively. The favoured simple fission

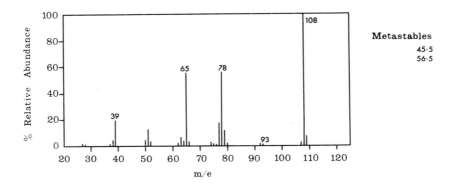

process is α-fission. (1) is ruled out because this would yield a peak of m/e 91. (2) is possible and (3), (4) and (5) are ruled out because each of these would lose a hydrogen atom by α-fission. Further consideration of anisole suggests that the next step in the degradation sequence would be elimination of a molecule of CO from m/e 93 to yield the stable cyclopentadienyl ion. m/e 65 is a very abundant peak and a metastable peak confirms the transition. With the molecular ion another possibility for degradation is elimination of a neutral molecule of formaldehyde by means of a four-membered transition state to yield an ion of m/e 78. In addition, cleavage α to the aromatic ring yields the ion m/e 77. This correlation which is shown over page confirms that the unknown molecule is anisole.

104 INTERPRETATION OF THE MASS SPECTRUM

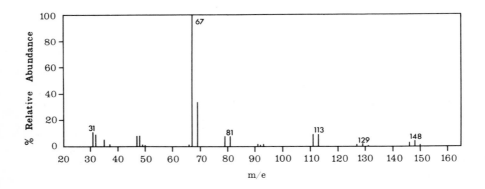

Example 5

There is a characteristic group of isotope peaks spaced two mass units apart. These peaks are at m/e 146, m/e 148 and m/e 150. The relative abundance ratio of these peaks is 3:4:1. The molecule must, therefore, contain one atom of chlorine and one atom of bromine. The presence of chlorine is confirmed by examination of the peaks at m/e 35 and m/e 37 with the abundance ratio of 3:1. Bromine gives rise to the peaks at m/e 79 and m/e 81 in the abundance ratio of 1:1. It can also be seen from the mass spectrum that the group of peaks m/e 127, m/e 129 and m/e 131 have an abundance ratio of 3:4:1. The mass difference between the ion m/e 146 and m/e 127 is 19, corresponding to a neutral fragment which does not have an effect on the isotope peak pattern. The common fragment of mass 19 is fluorine. Fluorine is mono-isotopic. The sum of the masses of fluorine, chlorine and bromine accounts for 133 mass units (19 + 35 + 79), which leaves 13 mass units to be accounted for. This mass corresponds to one hydrogen atom plus one carbon atom. The peaks at m/e 31 and m/e 32 correspond to CF and CHF respectively. The corresponding peaks for $C^{35}Cl$, $C^{37}Cl$, $CH^{35}Cl$ and $CH^{37}Cl$ are also present together with the peaks for $C^{79}Br$, $C^{81}Br$, $CH^{79}Br$ and $CH^{81}Br$. Peaks are also present due to the loss of chlorine and bromine respectively from the molecular ion. The compound under examination is therefore fluorochlorobromomethane. The mono-isotopic molecular weight is 146, which corresponds to $^{12}C^1 H^{19} F^{35} Cl^{79} Br$.

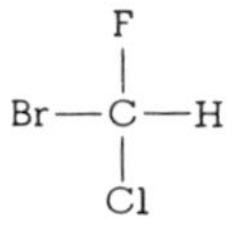

Example 6

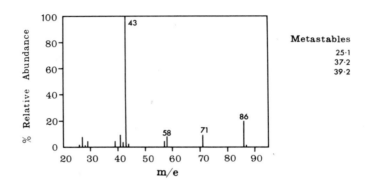

There are no characteristic isotope peaks and the molecular ion peak is at m/e 86. The formulae C_6H_{14}, $C_4H_6O_2$, $C_5H_{10}O$ and $C_4H_{10}N_2$ are possible and a wide range of possible compounds exists, including hydrocarbons, dialdehydes, diketones, diepoxides, unsaturated acids, esters and diamines. Exact mass measurement shows the molecular formula to be $C_5H_{10}O$. Possibilities include alcohols and ethers with one double bond equivalent, epoxides, aldehydes and ketones. Alcohols are ruled out because no loss of 17 (OH) or 18 (H_2O) mass units occurs and peaks in the homologous series m/e 31 + n × 14 do not occur. Ethers are ruled out because no peaks are observed in the series m/e 45 + n × 14 and a rearrangement peak at m/e 58 is not expected in the spectrum of an ether. Epoxides are also out because a double rearrangement peak in the series m/e 45 + n × 14 would be expected. This narrows the field down to aldehydes and ketones. Aldehydes generally have a $[M-1]^+$ ion and a peak at m/e 29 due to cleavage α to the carbonyl group.

The molecule is, therefore, a ketone and three possibilities exist:

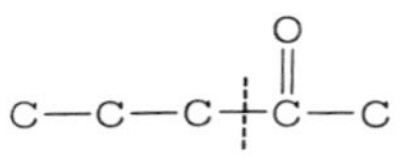

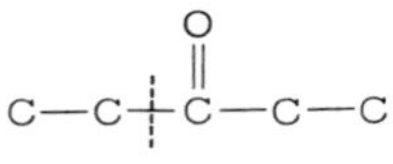

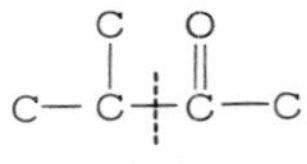

Simple fission to yield the base peak at m/e 43 eliminates structure (2), which would have a base peak at m/e 57. With ketones which are suitably substituted a McLafferty rearrangement is a favoured process. The rearrangement ion at m/e 58 corresponds to loss of an ethylene molecule from the molecular ion which can only be (1). The molecule is, therefore, methyl n-propyl ketone and the correlation is as follows:

$$CH_3CH_2CH_2-\overset{+}{\underset{\Vert}{C}}{=}O \quad \xleftarrow{2} \quad \left[CH_3CH_2CH_2\underset{2}{+}\overset{O}{\underset{\Vert}{C}}+CH_3 \right]^{+\cdot} \quad \xrightarrow{1} \quad \overset{+}{C_3H_7} \quad \text{and} \quad \overset{+}{\underset{\Vert}{C}}{=}O-CH_3$$

m/e 71   m/e 86   m/e 43   m/e 43

$$\downarrow$$

$$CH_3CH_2\overset{+}{C}H_2$$

m/e 43

m/e 86   ⟶   m/e 58   + $CH_2{=}CH_2$

QUALITATIVE AND QUANTITATIVE ANALYSIS OF MIXTURES
This should only be attempted if no other method is appropriate.

Qualitative analysis
In all cases, where possible, it is much better to separate a mixture before mass spectrometric analysis. If this is not possible, the following general procedure should be adopted.

1. Determine the spectrum of the mixture as for a pure compound. To avoid missing a less volatile component it is necessary to either evaporate the whole aliquot of the mixture into the appropriate reservoir system or allow some portion of this aliquot in, then carry out the procedure outlined here, pump the portion away and repeat with the next portion of the aliquot. The use of a direct insertion system should be avoided, if possible, since due to different volatilities one component of the mixture may disappear before the end of the determination.
2. Since ionization potentials tend to be lower than appearance potentials (see p. 28), if the spectra are run at low enough electron voltages the molecular ions will become relatively more important and can thus be identified. If the spectra can be displayed visually, then the correct electron voltages can be determined by inspection. If not, a series of spectra with electron voltages ranging from 15 to 7 eV should be determined.
3. Mass measurement will give the ionic formulae as usual and, if the molecular ions have been identified, the molecular formulae will be known.

4. Further analysis of the spectrum is normally difficult unless the use of metastables (see p. 23) allows the assignment of the major fragmentation peaks to their respective molecular ions. Sometimes the ionic formulae may help also, i.e. a molecular ion which has no nitrogen atoms cannot give rise to a fragment ion which has. If this assignment is possible, an attempt can be made to determine the structure of each component by the methods outlined on p. 91.

Quantitative analysis

This can only be done if pure samples of the components are available. The spectra of each of these should first be determined under the same conditions as are to be used for the mixture. The relative intensity of a peak or peaks is then used as an indication of the amount of each component present. The peaks chosen for one such component should preferably only appear in the spectrum of that component, otherwise an analysis similar to that of isotopically labelled molecules (see p. 30) will be necessary. A standard mixture is then run under the conditions to be used for analysis. It is essential that all the sample is evaporated into the sample reservoir. The peaks chosen above have their intensity measured, and this is correlated with the known composition of the mixture. This procedure should be repeated with known mixtures of different composition to check the procedure.

Finally, the spectrum of the unknown mixture is determined and the composition determined by means of the calibration experiments described above. The whole procedure is similar to that used in the infrared or ultraviolet spectral analysis of mixtures.

APPLICATION OF MASS SPECTROMETRY TO PROBLEMS IN ORGANIC CHEMISTRY

In a book of this size it is not possible to outline all the applications of mass spectrometry to the solution of problems involving pure compounds or mixtures of compounds. The sources of information mentioned at the end of this Chapter will supply a great deal of the information of this kind which has already been accumulated. It is hoped that from this book a sufficient understanding of the basic principles involved will be obtained to decide whether or not mass spectrometry could be useful in tackling any new problem.

SOURCES OF INFORMATION AND DATA REGARDING MASS SPECTROMETRY
Complete reference mass spectra

There are several collections of detailed reference mass spectra available. The most common of these are as follows:

1. *Catalogue of Mass Spectral Data,* American Petroleum Institute Research Project 44, Thermodynamics Research Centre, Texas A & M University. (Distributed in Europe by Heyden & Son Ltd.).
2. *Uncertified Mass Spectra,* circulated by sub-committee IV, ASTM Committee E.14.
3. *Collection of Uncertified Mass Spectra,* Dow Chemical Co., Chemical Physics Laboratory, Midland, Michigan.
4. *Mass Spectral Data Sheets,* MSDC, AWRE, Aldermaston, Reading, England. This continuing collection of uncertified reference spectra now numbers 5000 which are available with a cumulative index in data sheet form and over 11000 reference spectra are available on magnetic tape.

Simplified mass spectra

In structural elucidation work only the eight largest peaks in a mass spectrum need to be used for reference purposes. Three collections of data summarized in this way are available:

1. *Index of Mass Spectral Data,* (AMD 11), American Society for Testing and Materials, 1969. (Distributed in Europe by Heyden & Son Ltd.).

   A comprehensive index of approximately 8000 mass spectra. The spectral data consist of the six strongest peaks in each spectrum and their relative intensities, along with a coded number and name of the compound.

2. *Compilation of Mass Spectral Data,* A. Cornu and R. Massot, Heyden & Son, London, 1966 and 1st and 2nd Supplements, 1967 and 1971.

   This collection provides a comprehensive index to approximately 7000 mass spectra. The spectral data consist of the ten strongest peaks in each spectrum and their relative intensities. The data are indexed in separate sections by:

   a) reference number
   b) molecular weight
   c) molecular formula
   d) fragment ion values

3. *Eight Peak Index of Mass Spectra Vols. I and II,* MSDC, AWRE, Aldermaston, Reading, England, 1970.

   This consists of the intensities of the eight most abundant peaks from over 17000 mass spectra, arranged in three tables according to:

   a) molecular weight
   b) molecular formula
   c) ascending order of m/e values

Ionization and appearance potential data

These data have become increasingly more important in the detailed interpretation of mass spectral processes. Two compilations exist:

1. *Tables of Ionization Potentials,* R. W. Kiser, U.S. Atomic Energy Commission, Office of Technical Information Report No. TID-6142, 1960 and Supplement, 1962.

2. *Ionization Potentials, Appearance Potentials and Heats of Formation of Gaseous Positive Ions,* J. L. Franklin, J. G. Dillard, H. M. Rosenstock, J. T. Herron, K. Draxl and F. H. Field, NSRDS–NBS, Washington, 1969.

   This covers the literature from 1955 to June 1966.

Bibliographies

There are thousands of papers dealing with mass spectrometry and related topics. The rate of growth is approximately 7000 a year. Most mass spectroscopists have a printed survey of this literature which is almost up-to-date. If the use of mass spectrometry is being contemplated, the resident mass spectroscopist should be consulted. If he does not have the necessary information at hand to discuss the problem he can suggest where to start looking by making use of his files. This saves time, a factor which is becoming increasingly important in research. Alternatively, the usual literature searching methods can be used, in particular the series shown on the next page:

1. *Mass Spectrometry Bulletin,* MSDC, AWRE, Aldermaston, Reading, England.
   This is a monthly current awareness guide to the literature of mass spectrometry and allied topics. References are located by visual scanning of over 250 current scientific journals, together with a number of abstract journals, reports lists and books, and by computer searching of more than 2500 scientific journals for relevant articles. Entries are recorded under eight sub-headings and detailed indexing is provided. Over 3000 compounds are reported annually by molecular weight, molecular formulae and chemical classification.
2. *Organic Mass Spectrometry, Cumulative Chemical Compound Index, 1968–71,* Heyden and Son, London, 1971.
   This provides references to all compounds mentioned in the journal.
3. *Index and Bibliography of Mass Spectrometry, 1963–65,* F. W. McLafferty and J. Pinzelik, Wiley, 1967.
4. *Structure Indexed Literature of Organic Mass Spectra,* edited by Society of Mass Spectroscopy of Japan, Academic Press of Japan, Vol. 1, 1966; Vol. 2, 1967.

Textbooks

There are several textbooks available on the various aspects of mass spectrometry. Those which are of great value to the organic chemist are listed first, together with a brief description of their contents. The rest are referred to in order to complete the coverage.

1. *Mass Spectrometry and its Applications to Organic Chemistry,* J. H. Beynon, Elsevier, Amsterdam, 1960.
   This textbook deals with all aspects of the subject in detail and is invaluable in a laboratory where experimental mass spectrometry is being carried out. The general survey of the literature from which material has been drawn, dealing with all aspects of mass spectrometry up to 1960, is impressive. In addition, a great amount of unpublished knowledge from the experience of the author and his co-workers has been made available.
2. *The Mass Spectra of Organic Molecules,* J. H. Beynon, R. A. Saunders and A. E. Williams, Elsevier, Amsterdam, 1968.
   This is a partial replacement for the above book and brings the chapters on the fragmentation of organic molecules more up-to-date. It does not deal with the background to the subject in such detail as the earlier book.
3. *Mass Spectrometry – Organic Chemical Applications,* K. Biemann, McGraw-Hill, New York, 1963.
   This textbook deals with mass spectrometry from the viewpoint of the organic chemist. The fragmentation behaviour of the positive ions formed by electron bombardment is discussed in terms of the stability concepts of carbonium ions with which the chemist is familiar. Fragmentation rules which apply when specific functional groups or structural features are present in an ion are developed during these arguments. Some of the other sections of great interest include those on isotope labelling and applications of mass spectrometry to the structural elucidation of natural products.
4. *Structural Elucidation of Natural Products by Mass Spectrometry,* H. Budzikiewicz, C. Djerassi and D. H. Williams, Holden-Day, San Francisco, 1964.
   Volume 1 – Alkaloids
   This volume presents the discussion of the fragmentation behaviour of many classes of alkaloids using the approach developed in the book by these authors shown below. A useful section on isotope labelling and its applications is included.

Volume 2 — Steroids, Triterpenes and Related Classes
This volume extends the discussion to include an extensive selection of classes of compounds which have been studied by mass spectrometry. In addition, a set of tables by J. Lederberg is included which enables the atomic constitution of any mass measured ion to be easily found.

5. *Mass Spectrometry of Organic Compounds,* H. Budzikiewicz, C. Djerassi and D. H. Williams, Holden-Day, San Francisco, 1967.
This book discusses the fragmentation behaviour of the molecular ions and fragment ions of a wide range of classes of organic compounds. Mechanisms for these fragmentation processes are proposed and whenever possible isotope labelling evidence is put forward in their support. Other information such as that on metastable peaks and exact measurements is included where appropriate.

6. *Mass Spectrometry of Organic Ions,* F. W. McLafferty (Ed.), Academic Press, New York and London, 1963.
This contains the following chapters which have been written by experts in many branches of the subject, including the editor, whose chapter should be read by readers wishing to become proficient in the interpretation of mass spectra.

Quasi-Equilibrium Theory of Mass Spectrometry, H. M. Rosenstock and M. Kraus.
Ion—Molecule Reactions, C. E. Melton.
Appearance Potential Data of Organic Molecules, M. Kraus and V. H. Dibeler.
Negative Ion Mass Spectrometry, C. E. Melton.
Mass Spectrometry of Organic Radicals, A. G. Harrison.
Mass Spectrometry of Ions from Electric Discharges, Flames and Other Sources, P. F. Knewstubb.
Decompositions and Rearrangements of Organic Ions, F. W. McLafferty.
High Resolution Mass Spectrometry, R. A. Saunders and A. E. Williams.
Mass Spectrometry of Long Chain Esters, R. Ryhage and E. Stenhagen.
Mass Spectrometry of Alkyl Benzenes, H. M. Grubb and S. Meyerson.
Application to Natural Products and Other Problems in Organic Chemistry, K. Biemann
The Molecular Structure of Petroleum, A. Hood.
Mass Spectrometry of Terpenes, R. I. Reed.

Other books
Mass Spectra and Isotopes, F. W. Aston, Edward Arnold, London, 1942.
Mass Spectrometer Researches, G. P. Barnard, H. M. Stationery Office, London, 1956.
Mass and Abundance Tables for Use in Mass Spectrometry, J. H. Beynon and A. E. Williams, Elsevier, Amsterdam, 1963.
Tables for Use in High Resolution Mass Spectrometry, R. Binks, J. S. Littler and R. L. Cleaver Heyden and Son, London 1970.
Mass Spectroscopy, H. E. Duckworth, Cambridge University Press, London, 1958.
Electron Impact Phenomena and the Properties of Gaseous Ions, F. H. Field and J. L. Franklin, Academic Press, New York, 1971.
Introduction to Mass Spectrometry and its Applications, R. W. Kiser, Prentice Hall, New Jersey, 1965.
Computation of Molecular Formulae for Mass Spectrometry, J. Lederberg, Holden-Day, San Francisco, 1964.

Electronic and Ionic Impact Phenomena, H. S. W. Massey and E. H. S. Burlop, Oxford University Press, New York, 1969.
Mass Spectrometry, C. McDowell (Ed.), McGraw-Hill, New York, 1964.
Mass Spectral Correlations, F. W. McLafferty, American Chemical Society, Washington DC, 1963.
The Interpretation of Mass Spectra, F. W. McLafferty, Benjamin, New York, 1967.
The Principles of Mass Spectrometry and Negative Ions, C. E. Melton, Marcel Dekker, New York, 1970.
Determination of Organic Structures by Physical Methods, F. C. Nachod and W. D. Phillips (Eds.), Academic Press, New York, 1962.
Dynamic Mass Spectrometry, Vol. 1 – Proceedings of the 2nd European Time-of-Flight Mass Spectrometer Symposium, 1969, D. Price and J. E. Williams (Eds.), Heyden and Son, London, 1970.
Dynamic Mass Spectrometry, Vol. 2 – Invited papers and commissioned reviews, D. Price (Ed.), Heyden and Son, London, 1971.
Dynamic Mass Spectrometry, Vol. 3 – Proceedings of the 3rd International Time-of-Flight Mass Spectrometry Symposium, 1971, D. Price (Ed.), Heyden and Son, London (to be published 1972).
Mass Spectrometry, A. J. B. Robertson, Methuen, London, 1954.
Ion Production by Electron Impact, R. I. Reed, Academic Press, London, 1962.
Introduction to Mass Spectrometry, S. R. Schrader, Allyn and Bacon, Boston, 1971.
Advances in Mass Spectrometry, Vols. 1–5, (various editors), Pergamon Press, London, 1959–1971.

Journals

OMS–Organic Mass Spectrometry, An International Journal, A. Maccoll (Editor-in-Chief), Heyden and Son, London, from 1968. (Regional editors in eight countries, monthly).
International Journal of Mass Spectrometry and Ion Physics, J. Franzen, A. Quayle and H. J. Svec, (Eds.), Elsevier, Amsterdam, from 1968. (Approximately 10 issues a year).

INDEX TO UNKNOWN SPECTRA

1. 2-methylbutanol
2. 2-pentanethiol
3. *t*-butylamine
4. n-butyl methyl ether
5. methyl n-butyl thioether
6. methyl isobutyl amine
7. methyl *t*-butyl ketone
8. n-pentanal
9. pivalic acid
10. ethyl propionate
11. n-pentanamide
12. 4-methyl cyclohexanol
13. 1,4-dioxaspiro(4,5)decane
14. N,N'-dimethylcyclohexylamine
15. 1-fluorobutane
16. 1,2-dichloroethane
17. chlorobromomethane
18. iodobenzene

References

1. J. H. Beynon, *Mass Spectrometry and its Applications to Organic Chemistry*, Elsevier, Amsterdam, 1960.
2. (a) J. H. Beynon and A. E. Williams, *Mass and Abundance Tables for Use in Mass Spectroscopy*, Elsevier, Amsterdam, 1963; (b) J. Lederberg, *Compilation of Molecular Formulae for Mass Spectrometry*, Holden-Day Inc., San Francisco, 1964.
3. F. H. Field and J. L. Franklin, *Electron Impact Phenomena and the Properties of Gaseous Ions*, Academic Press, New York, 1971.
4. H. Budziekiewicz, C. Djerassi and D. H. Williams, *Structural Elucidation of Natural Products by Mass Spectrometry, Volume I – Alkaloids*, Holden-Day Inc., San Francisco, 1964.
5. K. Biemann, *Mass Spectrometry. Organic Chemical Applications*, McGraw-Hill, New York, 1962.
6. H. Budziekiewicz, C. Djerassi and D. H. Williams, *Interpretation of Mass Spectra of Organic Compounds*, Holden-Day Inc., San Francisco, 1967.
7. H. Budziekiewicz, C. Djerassi and D. H. Williams, *Structural Elucidation of Natural Products by Mass Spectrometry, Volume II – Steroids, Triterpenes and Related Classes*, Holden-Day Inc., San Francisco, 1964.
8. F. W. McLafferty, *Mass Spectrometry of Organic Ions*, Academic Press, London, 1963.
9. R. G. Cooks, 'Bond Formation in Organic Ions', *Org. Mass Spectrom.* **2**, 481 (1969).

Index

INTRODUCTION

Points common to a particular class of compounds are listed only under the general class heading. Individual compounds are listed under the general class heading after the general references. All general references to oxygen compounds also discuss the corresponding sulphur compounds and the individual sulphur compounds will be found under the corresponding class of oxygen compounds. A reference of the type 72/1 means that the spectrum of this compound can be found on p. 72 under the heading 'Unknown spectrum 1'. A reference of the type 93(f) means that a summary of the fragmentation patterns of these compounds is given on p. 93. Subjects only mentioned once and included in the Table of Contents have not been included in the index.

Acetals—
    diethyl, 69
    diisobutyl, 44
Acids, 78, 94(f)
    n-butanoic, 99
    o-methylbenzoic, 90
    pivalic, 81/9
    n-pentanoic, 78
Alcohols, 34, 38, 40, 43, 46, 52, 58, 70, 93(f)
    cyclic, 34, 38, 53, 62, 82
    loss of H_2O, H_2S, 48, 53, 71, 83
    α-cleavage and rearrangement, 58, 72, 100
    trimethylsilyl derivatives, 33
    butanols—
        isobutanol, 58
        n-butanol, 70
        2-methyl, 72/1
        s-butanol, 41
    cyclohexanol, 53, 62, 82
        4-methyl, 84/12
        monomethyl, 38
        thiol, 82

Alcohols—*contd.*
    ethanol, 28
    hexan-3-ol, 59
    n-pentanethiol, 71
    2-pentanethiol, 72/2
    pentan-2-ol, 99
    β-phenyl ethyl, 89
Aldehydes, 38, 44, 48, 50, 68, 76, 94(f)
    aromatic, 95(f)
    acetaldehyde, 44
    2-methylbutanal, 77
    n-pentanal, 78/8
Alkaloids, 96
    crotonosines, 66
    ibogamine, 97
    ibogaine, 58, 97
    ibogaline, 97
    tabernanthine, 97
Amides, 33, 50, 78, 94(f)
    n-butanamide, 79
    n-pentanamide, 81/11
Amines, 33, 38, 41, 43, 52, 94(f)
    α-cleavage and rearrangement, 58, 59, 72, 74, 102

Amines—*contd.*
    cyclic, 62, 82
    primary, 70
    secondary and tertiary, 73
    aniline, 90
    n-butyl, 71
    t-butyl, 73/3
    cyclohexyl, 62, 82, 83
        N-dimethyl, 84/14
    ethyl, 28
    ethyl n-propyl, 74
    methyl isobutyl, 76/6
    methyl diethyl, 100
    pyrrole, 48
    steroidal dimethyl, 33, 63, 64
    tetrahydroisoquinoline, 90
    triethyl, 60
Amino acids, 30
Amino alcohols—
    ethanolamine, 28, 41, 43
    1-hydroxy-2-aminoisobutane, 43
Anthraquinone, 91
Applications, 1, 3, 91, 106, 108
Aromatic compounds, 46, 50, 51, 67, 68, 88, 95(f)
    alkyl, 47, 50, 88, 95(f)
    ortho effects, 52, 90
    sulphonyl chlorides, 61

Carbonates, 54
    di-n-propyl, 55
Charge localization, 36, 38, 41
Computers—
    data processing, 9, 12, 13, 25
    isotope patterns calculation, 17
    mass measurement, 25, 27
    metastables, 20
Cyclic compounds, 18, 38, 39, 53, 62, 69, 82, 95(f)

Electrostatic analyser, 7
Element map, 25
Epoxides, 50, 57
    1,2-epoxybutane, 57
    1,2-epoxypentane, 50
Esters, 33, 45, 50, 51, 53, 54, 68, 78, 95(f)
    acetates, 55
    β-phenyl ethyl, 52
    unsaturated, 36
    benzyl acetate, 54
    n-butyl furoate, 56
    2-chloroallyl acetate, 56
    cyanomalonates, 61
    diethyl malonate, 56
    ethyl butanoate, 29
    ethyl propionate, 81/10

Esters—*contd.*
    β-hydroxy-*p*-anisate, 55
    isopropyl n-butanoate, 56
    methyl n-butanoate, 79
    methyl propionate, 45
Ethers, 33, 38, 40, 42, 43, 46, 94(f)
    aliphatic, 73
    aromatic, 46, 80, 88, 89
    α-cleavage and rearrangement, 59, 74
    cyclic, 62, 82
    polyethers, 61
    anisole, 90, 103
    n-butyl methyl, 75/4
    n-butyl methyl thioether, 76/5
    2-chloroethyl phenyl, 89
    chroman, 91
    diethyl, 73
    ethyl phenyl, 89
    ethyl n-propyl thioether, 73
    isochroman, 66
    monomethyl of glycols, 42
Ethylene ketals, 33, 36, 63, 82
    of cyclohexanone, 84/13
    of steroids, 63

Field-free regions, 7
Fragmentation (*see also* Rearrangements)
    classifications, 39
    α-, β-, γ-cleavages, definitions, 37
    derivatives, use of, 33
    Hammett constants, 39
    ionization potentials, 36, 41
    ring ions, 37, 87
    simple, 39
    Stevenson's rule, 28, 41
    types of, 39–67
Furan, 48

Gas chromatography, 3, 9
Glycols, 34, 42, 56
    ethylene, 42
    propylene, 42
    2-methoxy-*s*-butanol, 42
    2-methoxy-*t*-butanol, 42
    2-methoxyethanol, 42, 57
    1-methoxyisopropanol, 42

Halides, 37, 40, 43, 85, 95(f)
    aromatic, 89
    loss of hydrogen halides, 53, 86
    bromoethane, 85
    t-butyl chloride, 36
    n-butyl fluoride, 87/15
    chlorobromomethane, 87/17
    chloroethane, 85
    1,2-dibromobenzene, 68

Halides—*contd.*
    1,2-dichloroethane, 87/16
    fluorochlorobromomethane, 104
    fluoroethane, 85
    hexachlorobutadiene, 11
    iodobenzene, 88/18
    iodoethane, 86
Hydrocarbons, 4, 39, 48, 93(f)
    ethane, 28
    3-ethyl-3-methylpentane, 67
    n-octane, 22, 40, 68
        1-phenyl, 50

Ion formulae determination—
    (*see Molecular formulae determination*)
Ion intensity recording, 8
    computer, 9, 12
    display, 12–14
    electrical, 5, 8, 9, 10, 13
    photoplate, 5, 8, 9, 10
Ionization—
    chemical, 4
    efficiency, 4
    field, 4
    photons, 5
    potential, 28, 36, 41
        of acetone, 37
        of ethane, 28
        of ethanol, 28
        of ethylamine, 28
        fragmentation and, 48, 51
Ion—molecule reactions, 4, 18, 19, 92
Isotopes, 1, 14
    calculation of ion intensities, 15
    labelling, 23, 28, 29
        determination of amount of, 30, 85
    molecular formulae determination, 27
    natural abundances, 14
    uses, 29, 47, 104

Ketones, 38, 44, 48, 68, 76, 94(f)
    aromatic, 46, 88
    cyclic, 39, 82
    McLafferty rearrangement, 50
    acetone, 36, 37
    acetophenone, 11, 20, 21
    n-butyl isobutyl, 51
    cyclohexanone, 62, 82, 84
        monomethyl compounds, 39
    cyclopentanone, 83
    decal-2-one-1-ene, 64
    2,2-diethylcyclohexanone, 52
    2,6-diethylcyclohexanone, 52
    methyl isobutyl, 76
    methyl *t*-butyl, 78/7
    methyl n-propyl, 45, 50, 105

Ketones—*contd.*
    3-nitro-4-chloro acetophenone, 68
    n-propyl n-butyl, 51

Mattauch–Herzog, 7
McLafferty rearrangement, 36, 49, 50
    acids, 80, 99
    alcohols, 100
    amides, 80
    aromatics, 51, 89
    carbonyl compounds, 77, 105
    esters, 29, 54, 80
    extent of, 50
    factors governing, 49, 50, 52
    Stevenson's rule, 51
Metastables, 11, 12
    applications, 23, 67, 68, 92
    assignment of *m/e* values, 11
    detection in first field-free region, 22
    examples, 11, 12
    labelling, 23
    parent—daughter ion determination, 19
    reason for, 19
Molecular ions—
    formation of, 17
    identification of, 4, 18, 19, 92
    nitrogen rule, 36
    stability of, 18, 35
Molecular formulae determination, 23, 27, 33, 34
    accurate mass measurement, 25
    chemical pretreatment, 32
    computer, 25, 27
    ion-molecule reactions, 4, 18, 19
    isotopes, 27
    nitrogen rule, 27
    peak matching, 26
Multiply-charged ions, 12, 19

Naphthalene, 35
Nier–Johnson, 7
Nitriles, 50, 53, 95(f)
    ion-molecule reactions, 19
    n-propyl, 50
Nitrobenzene, 88, 89
Nitrogen rule, 36
    cyclic systems, 95
    fragment ions, 36
    molecular formulae, 27, 36

Olefines, 45, 50, 51, 93(f)
    cis/trans isomers, 52
    cyclic, 64
    location of double bond, 34
    molecular ions, 38
    cyclohexene, 64

Olefines—*contd.*
    2,3-dimethylcyclohexene, 65
    3-methyl-*trans*-2-pentene, 20
    Δ^{12}-oleanene, 65
    pentacyclic triterpenes, 65
Ortho effects, 52, 90

Peptides, 30
Phenols, 52, 89
    phenol, 91
        o-hydroxybenzyl, 90
Phenylisothiocyanate, 89
Phenylnitrite, 46
1-phenyloctane, 50
Phenylthioureas, 37
Pyridine, alkyl derivatives, 46
Pyrolysis, 4, 35, 53, 71
    alcohols, 46, 53, 71
    chemical pretreatment, 32
    sample inlets, 3

Rearrangements, 48
    aromatics, 51, 67
    classification, 48
    complex, 62, 83
    in fragment ions, 58, 67
    governing factors, 49
    hydrogen—
        double, 54, 80, 83
        random, 49, 51, 53, 64, 71
        specific (*see also McLafferty*),
        45, 49, 74

Rearrangements—*contd.*
    molecules eliminated by, 53, 58, 67
    multi-step, 62
    radicals other than hydrogen, 48, 60
Resolution, 1, 7, 24
    definition, 24
    limiting factors, 24
Retro-Diels–Alder, 49, 64
    applications, 65
    aromatics, 66, 90

Sample—
    inlets, 2
    pretreatment, 33
    size, 3
Stevenson's rule, 28, 41
    McLafferty rearrangement, 51
Structure determination—
    derivatives and, 33
    examples of, 98
    general procedure, 91, 96
Sulphonamides, 90

Tetralin, 49
Thiophene, 48
    2-benzoyl, 61
Toluene (*see Tropylium*)
Tropylium ion, 47, 58
    ionization potential, 28

Compilation of Mass Spectral Data

PART D

By Fragment Ion Values

A. Cornu & R. Massot

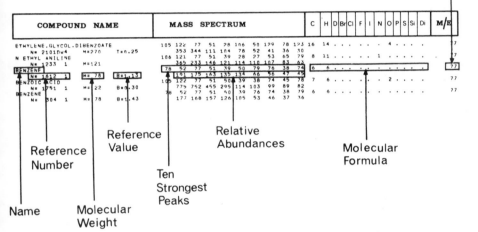

7000 spectra! available in one main volume with two supplements. The compounds are sorted into four sections and are tabulated by: Reference Number; Molecular Weight; Molecular Formula and Fragment Ion Values. The tables also detail the origin, reference value and ten strongest peaks, with relative abundances, of the compounds.

All the data which make up this compilation is obtainable in the form of approximately 14000 IBM punched cards and magnetic tape; a retrieval programme has also been written — details available on request.

Complete Set £22.50 $54.00 DM 203.00

Spectrum House   Alderton Crescent   London NW4 3XX   01-202 5333

DYNAMIC MASS SPECTROMETRY

VOLUME 1 (Edited by D. Price & J. E. Williams) "can be recommended as an essential ...... because it covers a wide range of interests" *Laboratory Practice* (1969). A compilation of the most significant papers presented at the 2nd European Time of Flight Mass Spectrometry Symposium, July 1969. The four main sections cover Kinetics Studies, Ionization Processes. Analytical Applications and Instrumentation.
247 pp. £5.50 $13.50 DM 50.00

Edited by D. Price and published in 1971. Contains specially commissioned reviews and papers on Ion Cyclotron Resonance; Computerised GC/Quadrupole MS; Laser Dynamic Mass Spectrometry; Fast Gas Phase Reactions; Negative Ion Lifetimes; etc. An all-embracing bibliography enhances the value of this volume. **VOLUME 2**
272 pp £6.25 $16.00 DM 54.00

VOLUME 3 (Edited by D. Price). Published in Autumn 1972. Features specifically invited reviews on Applications of Inhomogeneous Oscillatory Electric Fields in Ion Physics; Negative Ions and Residual Gas Analysis. The remainder of the book consists of the major papers delivered at the 3rd European Time of Flight Symposium, July 1971 together with a bibliography that updates that of Volume 2.

HEYDEN & SON LTD

Spectrum House   Alderton Crescent   London NW4 3XX   01-202 5333

Speed. Efficiency. Accuracy. The main criteria for working in today's busy laboratories. And this complete set of tables is designed to meet just these criteria.

They will provide vital assistance in instrument calibration and rapid determination of the exact masses of ions. The accompanying 32-page book of tables, **Chemical Formulae from Mass Determinations**, by D. Hennenberg and K. Casper, enables the reader to convert these exact masses into chemical formulae. The explanatory text, throughout, is in English, French and German.

tables for use in high resolution mass spectrometry

R. Binks, J. S. Littler & R. L. Cleaver

CONTENTS

Table 1 Formulae, exact masses and carbon-13 ratios of all possible combinations of 1-20 carbon, 1-39 fluorine and 0 or 1 nitrogen atoms.

Table 2 The ratios between each useful reference peak (up to 1.5) and all the other possible C, F and N combinations considered in Table 1.

Table 3 As for Table 2 but for fluothane only.

Table 4 The ratios between any peak and that one mass unit higher.

Table 5 Peak patterns (plotted graphically and tabulated numerically) to be expected due to the presence of chlorine, bromine, sulphur, silicon, boron, and bromine/chlorine up to a maximum of ten heteroatoms.

Table 6 Data used in table preparation.

180 pp. (plus 32 pp. booklet) £10.80 $26.00 DM 97.00

Spectrum House   Alderton Crescent   London NW4 3XX   01-202 5333

ORGANIC MASS SPECTROMETRY
An International Journal

Designed for and compiled by mass spectrometrists with an organic bias. This widely acclaimed monthly journal is the international forum for rapid publication of papers, exchanges of ideas and short communications in English, French or German. Each issue carries a useful cumulative Compound Index in supplementary form. A Keyword Subject Index, which greatly enhances the value of the journal, appears from late 1972.

EDITOR-IN-CHIEF
PROF. ALLAN MACCOLL

Assistant Editor:
DR. A. G. LOUDON
University College, 20 Gordon Street, London WC1H 0AJ

REGIONAL EDITORS
NORTH AMERICA
Prof. Robert H. Shapiro, *Department of Chemistry, University of Colorado, Boulder, Colorado 80302*

Associate Editor:
Dr. S. P. Markey, *Department of Pediatrics, University of Colorado Medical Center, Denver, Colorado 80220*

FRANCE
Prof. A. Cornu, *Commissariat a l'Énergie Atomique, Centre d'Études Nucléaires, Cedex No. 85, Grenoble 38*

GERMANY
Prof. H.-Fr. Grützmacher, *Institut für Organische Chemie der Universität, 2 Hamburg 13, Papendamm 6*

NETHERLANDS
Prof. G. Dijkstra, *Analytisch Chemisch Laboratorium der Rijksuniversiteit Utrecht, Croesestraat 77a, Utrecht*

USSR
Prof. N. K. Kochetkov, *N.D. Zelinsky Institute of Organic Chemistry, Academy of Sciences of the USSR, Moscow B 334, Leninsky prospect 47*

AUSTRALASIA
Prof. J. S. Shannon, *Chemistry School, University of New South Wales, P.O. Box 1, Kensington, N.S.W. 2033, Australia*

JAPAN
Prof. Akira Tatematsu, *Faculty of Pharmacy, Meijo University, Showa-ku, Nagoya*

One Volume per annum — 12 Issues per Volume
Annual Subscription   £35.00 $85.00 DM 310.00

HEYDEN & SON LTD

Spectrum House   Alderton Crescent   London NW4 3XX   01-202 5333